Introduction to 3-Manifolds

Jennifer Schultens

Graduate Studies
in Mathematics
Volume 151

BOWLING GREEN STATE
UNIVERSITY LIBRARIES
American Mathematical Society
Providence, Rhode Island

EDITORIAL COMMITTEE

David Cox (Chair)
Daniel S. Freed
Rafe Mazzeo
Gigliola Staffilani

2010 *Mathematics Subject Classification.* Primary 57N05, 57N10, 57N16, 57N40, 57N50, 57N75, 57Q15, 57Q25, 57Q40, 57Q45.

For additional information and updates on this book, visit
www.ams.org/bookpages/gsm-151

Library of Congress Cataloging-in-Publication Data
Schultens, Jennifer, 1965–
  Introduction to 3-manifolds / Jennifer Schultens.
    pages cm — (Graduate studies in mathematics ; v. 151)
  Includes bibliographical references and index.
  ISBN 978-1-4704-1020-9 (alk. paper)
  1. Topological manifolds. 2. Manifolds (Mathematics) I. Title. II. Title: Introduction to three-manifolds.

QA613.2.S35 2014
514′.34—dc23
                                                            2013046541

Copying and reprinting. Individual readers of this publication, and nonprofit libraries acting for them, are permitted to make fair use of the material, such as to copy a chapter for use in teaching or research. Permission is granted to quote brief passages from this publication in reviews, provided the customary acknowledgment of the source is given.

Republication, systematic copying, or multiple reproduction of any material in this publication is permitted only under license from the American Mathematical Society. Requests for such permission should be addressed to the Acquisitions Department, American Mathematical Society, 201 Charles Street, Providence, Rhode Island 02904-2294 USA. Requests can also be made by e-mail to reprint-permission@ams.org.

© 2014 by the author.
Printed in the United States of America.

∞ The paper used in this book is acid-free and falls within the guidelines established to ensure permanence and durability.
Visit the AMS home page at http://www.ams.org/

10 9 8 7 6 5 4 3 2 1   19 18 17 16 15 14

Dedicated to Misha and Esther and our extended family

Contents

Preface		ix
Chapter 1. Perspectives on Manifolds		1
§1.1.	Topological Manifolds	1
§1.2.	Differentiable Manifolds	7
§1.3.	Oriented Manifolds	10
§1.4.	Triangulated Manifolds	12
§1.5.	Geometric Manifolds	21
§1.6.	Connected Sums	23
§1.7.	Equivalence of Categories	25
Chapter 2. Surfaces		29
§2.1.	A Few Facts about 1-Manifolds	29
§2.2.	Classification of Surfaces	31
§2.3.	Decompositions of Surfaces	39
§2.4.	Covering Spaces and Branched Covering Spaces	41
§2.5.	Homotopy and Isotopy on Surfaces	45
§2.6.	The Mapping Class Group	47
Chapter 3. 3-Manifolds		55
§3.1.	Bundles	56
§3.2.	The Schönflies Theorem	62
§3.3.	3-Manifolds that are Prime but Reducible	71

§3.4.	Incompressible Surfaces	72
§3.5.	Dehn's Lemma*	75
§3.6.	Hierarchies*	80
§3.7.	Seifert Fibered Spaces	87
§3.8.	JSJ Decompositions	96
§3.9.	Compendium of Standard Arguments	98

Chapter 4.	Knots and Links in 3-Manifolds	101
§4.1.	Knots and Links	101
§4.2.	Reidemeister Moves	106
§4.3.	Basic Constructions	108
§4.4.	Knot Invariants	113
§4.5.	Zoology	118
§4.6.	Braids	122
§4.7.	The Alexander Polynomial	126
§4.8.	Knots and Height Functions	128
§4.9.	The Knot Group*	137
§4.10.	Covering Spaces*	139

Chapter 5.	Triangulated 3-Manifolds	143
§5.1.	Simplicial Complexes	143
§5.2.	Normal Surfaces	148
§5.3.	Diophantine Systems	155
§5.4.	2-Spheres*	162
§5.5.	Prime Decompositions	166
§5.6.	Recognition Algorithms	169
§5.7.	PL Minimal Surfaces**	172

Chapter 6.	Heegaard Splittings	175
§6.1.	Handle Decompositions	175
§6.2.	Heegaard Diagrams	180
§6.3.	Reducibility and Stabilization	182
§6.4.	Waldhausen's Theorem	188
§6.5.	Structural Theorems	193
§6.6.	The Rubinstein-Scharlemann Graphic	196
§6.7.	Weak Reducibility and Incompressible Surfaces	200

§6.8.	Generalized Heegaard Splittings	202
§6.9.	An Application	208
§6.10.	Heegaard Genus and Rank of Fundamental Group*	212
Chapter 7.	Further Topics	215
§7.1.	Basic Hyperbolic Geometry	215
§7.2.	Hyperbolic n-Manifolds**	220
§7.3.	Dehn Surgery I	226
§7.4.	Dehn Surgery II	232
§7.5.	Foliations	238
§7.6.	Laminations	243
§7.7.	The Curve Complex	248
§7.8.	Through the Looking Glass**	252
Appendix A.	General Position	261
Appendix B.	Morse Functions	269
Bibliography		275
Index		283

Preface

This book grew out of a graduate course on 3-manifolds taught at Emory University in the spring of 2003. It aims to introduce the beginning graduate student to central topics in the study of 3-manifolds. Prerequisites are kept to a minimum but do include some point set topology (see [**109**]) and some knowledge of general position (see [**128**]). In a few places, it is worth our while to mention results or proofs involving concepts from algebraic topology or differential geometry. This should not stop the interested reader with no background in algebraic topology or differential geometry from enjoying the material presented here. The sections and exercises involving algebraic topology are marked with a *, those involving differential geometry with a **.

This book conveys my personal path through the subject of 3-manifolds during a certain period of time (roughly 1990 to 2007). Marty Scharlemann deserves credit for setting me on this path. He remains a much appreciated guide. Other guides include Misha Kapovich, Andrew Casson, Rob Kirby, and my collaborators.

In Chapter 1 we introduce the notion of a manifold of arbitrary dimension and discuss several structures on manifolds. These structures may or may not exist on a given manifold. In addition, if a particular structure exists on a given manifold, it may or may not be presented as part of the information given. In Chapter 2 we consider manifolds of a particular dimension, namely 2-manifolds, also known as surfaces. Here we provide an overview of the classification of surfaces and discuss the mapping class group. Chapter 3 gives examples of 3-manifolds and standard techniques used to study 3-manifolds. In Chapter 4 we catch a glimpse of the interaction of pairs of manifolds, specifically pairs of the form (3-manifold, 1-manifold). Of particular interest here is the consideration of knots from the point of view of the complement ("Not Knot"). For other perspectives, we refer the reader to the many books, both new and old, mentioned in Chapter 4, that provide a more in-depth study. In Chapter 5 we consider triangulated 3-manifolds, normal surfaces, almost normal surfaces, and how these set the stage for algorithms pertaining to 3-manifolds. In Chapter 6 we cover a subject near and dear to the author's heart: Heegaard splittings. Heegaard splittings are decompositions of 3-manifolds into symmetric pieces. They can be thought of in many different ways. We discuss key examples, classical problems, and recent advances in the subject of Heegaard splittings. In Chapter 7 we introduce hyperbolic structures on manifolds and complexes and provide a glimpse of how they affect our understanding of 3-manifolds. We include two appendices: one on general position and one on Morse functions. Exercises appear at the end of most sections.

I wish to thank the many colleagues and students who have given me the opportunity to learn and teach. I also wish to thank the institutions that have supported me through the years: University of California, Emory University, Max-Planck-Institut für Mathematik Bonn, Max-Planck-Institut für Mathematik Leipzig, and the National Science Foundation.

Chapter 1

Perspectives on Manifolds

Manifolds are studied in a variety of categories. We will introduce several of these categories and highlight the advantages of the different points of view. We will focus on topological manifolds (TOP), differentiable (or smooth) manifolds (DIFF) and triangulated (or combinatorial) manifolds (TRIANG). To see these definitions in context, consult introductory Chapters in [93], [100], [48], [148], [63], [67], [59], or [37].

1.1. Topological Manifolds

Definition 1.1.1. A *topological n-manifold* is a second countable Hausdorff space M for which there exists a family of pairs $\{(M_\alpha, \phi_\alpha)\}$ with the following properties:

- $\forall \alpha$, M_α is an open subset of M and $M = \bigcup_\alpha M_\alpha$;
- $\forall \alpha$, ϕ_α is a homeomorphism from M_α to an open subset of $\mathbb{R}^n$.

We often refer to a topological manifold simply as a *manifold*.

A pair (M_α, ϕ_α) is called a *chart* of M. The family of pairs $\{(M_\alpha, \phi_\alpha)\}$ is called an *atlas* for M. (We sometimes write only $\{M_\alpha\}$ rather than $\{(M_\alpha, \phi_\alpha)\}$ when the maps ϕ_α are not part of the discussion.)

The *dimension* of an n-manifold is n. Two n-manifolds are considered equivalent if they are homeomorphic.

Remark 1.1.2. Every subset of $\mathbb{R}^N$ is second countable and Hausdorff. Thus to show that a subset M of $\mathbb{R}^N$ is an n-manifold it suffices to show that every point in M has a neighborhood homeomorphic to $\mathbb{R}^n$.

Requiring these homeomorphisms to map onto $\mathbb{R}^n$ is equivalent to requiring them to map onto open subsets of $\mathbb{R}^n$. In exhibiting homeomorphisms we often opt for the latter, in the interest of simplicity.

Example 1.1.3. One immediate example of an n-manifold is $\mathbb{R}^n$ and open subsets of $\mathbb{R}^n$.

Example 1.1.4. The set $\mathbb{S}^n = \{x \in \mathbb{R}^{n+1} : \|x\| = 1\}$ is an n-dimensional manifold called the *n-sphere*. Stereographic projection provides a homeomorphism $h : \mathbb{S}^n \backslash \{(0, \ldots, 0, 1)\} \to \mathbb{R}^n$. Thus any point $x \in \mathbb{S}^n$ such that $x \neq (0, \ldots, 0, 1)$ has the neighborhood $\mathbb{S}^n \backslash \{(0, \ldots, 0, 1)\}$ that is homeomorphic to $\mathbb{R}^n$. To exhibit a neighborhood of $(0, \ldots, 0, 1)$ that is homeomorphic to $\mathbb{R}^n$ we compose the reflection in $\mathbb{R}^n \times \{0\}$ with h to obtain $h' : \mathbb{S}^n \backslash \{(0, \ldots, 0, -1)\} \to \mathbb{R}^n$. Thus $\mathbb{S}^n \backslash \{(0, \ldots, 0, -1)\}$ is a neighborhood of $(0, \ldots, 0, 1)$ homeomorphic to $\mathbb{R}^n$. See Figure 1.1.

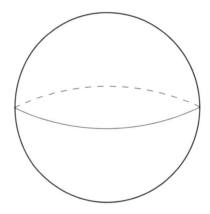

Figure 1.1. *The 2-dimensional sphere.*

Example 1.1.5. The set $\mathbb{T}^n = \mathbb{S}^1 \times \cdots \times \mathbb{S}^1$ (n factors) is called the *n-torus*. The n-torus is a quotient space obtained as follows: In $\mathbb{R}^n$, consider the group G generated by translations of distance 1 along the coordinate axes. Then identify two points $x, y \in \mathbb{R}^n$ if and only if there is a $g \in G$ such that $g(x) = y$. Denote this quotient map by $q : \mathbb{R}^n \to \mathbb{T}^n$.

To see that $\mathbb{T}^n$ is an n-manifold, let $[x] \in \mathbb{T}^n$ and let U be the sphere of radius $\frac{1}{4}$ centered at $x \in \mathbb{R}^n$. Note that $g(U) \cap U = \emptyset \ \forall g \in G$. It follows that $q^{-1}|_{q(U)} : q(U) \to U$ is a homeomorphism. The set of all such homeomorphisms provides an atlas for $\mathbb{R}^n$. See Figures 1.2 and 1.3.

Example 1.1.6. The result of identifying antipodal points on $\mathbb{S}^n$ is an n-manifold. It is called *n-dimensional real projective space* and is denoted by $\mathbb{R}P^n$. To verify that $\mathbb{R}P^n$ is an n-manifold, consider a point $[x]$ in $\mathbb{R}P^n$. Since

1.1. Topological Manifolds

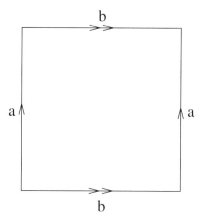

Figure 1.2. *The 2-dimensional torus is obtained from a square via identifications.*

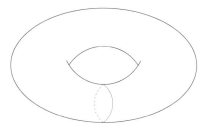

Figure 1.3. *The 2-dimensional torus.*

$\mathbb{S}^n$ is an n-manifold, there is a neighborhood U of x and a homeomorphism $h : U \to \mathbb{R}^n$. Set $-U = a(U)$, for $a : \mathbb{S}^n \to \mathbb{S}^n$ the antipodal map. Then $-U$ is a neighborhood of $-x$ and $-h = h \circ a$ is a homeomorphism between $-U$ and $\mathbb{R}^n$. After shrinking U, if necessary, $a(U) \cap U = \emptyset$. Thus we have a homeomorphism $[h] : [U] \to \mathbb{R}^n$. The set of all such homeomorphisms defines an atlas for $\mathbb{R}P^n$. See Figure 1.4.

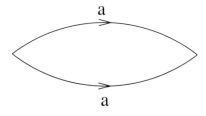

Figure 1.4. *The projective plane obtained from a bigon via identifications.*

Definition 1.1.7. Let M be an n-manifold. A *p-dimensional submanifold* of M is a closed subset L of M for which there exists an atlas $\{(M_\alpha, \phi_\alpha)\}$

of M such that $\forall x \in L$ there is a chart in the atlas with $x \in M_\alpha$ and
$$\phi_\alpha(L \cap M_\alpha) = \{\mathbf{0}\} \times \mathbb{R}^p \subset \mathbb{R}^n.$$

Remark 1.1.8. A submanifold is itself a manifold.

Example 1.1.9. The equatorial circle in the 2-sphere indicated in Figure 1.1 is a submanifold of the 2-sphere.

Definition 1.1.10. Let L, M be manifolds. A map $f : L \to M$ is an *embedding* if it is a homeomorphism onto its image $f(L)$ and $f(L)$ is a submanifold of M.

Example 1.1.11. If L is a submanifold of M, then the inclusion map $i : \tilde{L} \to M$ of an abstract copy $\tilde{L}$ of L to $L \subset M$ is an embedding.

We will also consider a slightly larger class of objects:

Definition 1.1.12. Set $H^n = \{(x_1, \ldots, x_n) \in \mathbb{R}^n : x_1 \geq 0\}$. An *n-manifold with boundary* is a second countable Hausdorff space M with an atlas such that $\forall \alpha$, ϕ_α is a homeomorphism from M_α to an open subset of $\mathbb{R}^n$ or H^n.

The *boundary* of M is the set of all points in M that have a neighborhood homeomorphic to H^n but no neighborhood homeomorphic to $\mathbb{R}^n$. The boundary of M is denoted by ∂M. Points not on the boundary are called interior points. Two n-manifolds with boundary are considered equivalent if they are homeomorphic.

Example 1.1.13. The set $\mathbb{B}^n = \{x \in \mathbb{R}^n : \|x\| \leq 1\}$ is an n-dimensional manifold with boundary called the *n-ball*. For interior points, there is nothing to check (because the identity map on $\mathbb{R}^n$ provides the required homeomorphism). For boundary points, an extension of the map obtained by stereographic projection provides the required homeomorphism. See Figure 1.5.

Figure 1.5. *The 2-ball is also called the disk.*

Example 1.1.14. The *pair of pants* is a 2-manifold with boundary. See Figure 1.6.

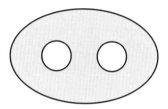

Figure 1.6. *The pair of pants.*

Example 1.1.15. The 1-holed torus is a 2-manifold with boundary. See Figure 1.7.

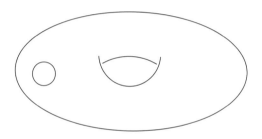

Figure 1.7. *The 1-holed torus.*

Definition 1.1.16. Let M be an n-manifold with boundary. A *p-dimensional submanifold* of M is a closed subset L of M for which there is an atlas $\{(M_\alpha, \phi_\alpha)\}$ of M and $p \in \{0, \ldots, n\}$ such that $\forall x \in L$ in the interior of M there is a chart in the atlas such that $x \in M_\alpha$ and

$$\phi_\alpha(L \cap M_\alpha) = \{\mathbf{0}\} \times \mathbb{R}^p \subset \mathbb{R}^n$$

and $\forall x \in L$ in the boundary of M there is a chart in the atlas such that $x \in M_\alpha$ and

$$\phi_\alpha(L \cap M_\alpha) = \{\mathbf{0}\} \times H^p \subset H^n$$

and such that

$$\phi_\alpha(x) \in \{\mathbf{0}\} \times \partial H^p \subset \partial H^n.$$

Remark 1.1.17. The boundary, ∂M, of an n-manifold M is not a submanifold of M, though it is an $(n-1)$-dimensional manifold that is contained in M.

Example 1.1.18. The diameter of the disk pictured in Figure 1.8 is a submanifold of a manifold with boundary.

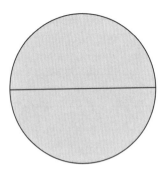

Figure 1.8. *A submanifold of the disk.*

Definition 1.1.19. We say that the n-manifold M is *closed* if M is compact and $\partial M = \emptyset$.

Example 1.1.20. Spheres and tori are examples of closed manifolds.

In the TOP category, that is, when we study manifolds from the point of view in this section, we are interested in continuous maps between manifolds.

Example 1.1.21. Projection from $\mathbb{T}^2 = \mathbb{S}^1 \times \mathbb{S}^1$ onto the second factor is a continuous map between manifolds.

To catch a glimpse of intriguing topological manifolds browse [78].

Exercises

Exercise 1. Convince yourself that the statements in Remark 1.1.2 are true.

Exercise 2. Prove that the product of two manifolds is a manifold. What can you say about its dimension? (This provides an alternate proof of the fact that $\mathbb{T}^n$ is a manifold.)

Exercise 3. Show that the boundary of an n-manifold with boundary is an $(n-1)$-manifold without boundary. See for instance the 1-holed torus in Figure 1.7.

Exercise 4. By drawing pictures, convince yourself that the surface with boundary called the pair of pants is aptly named.

Exercise 5. Argue that the 1-manifolds pictured in Figure 1.9 are equivalent. (You needn't give a formal proof.)

1.2. Differentiable Manifolds

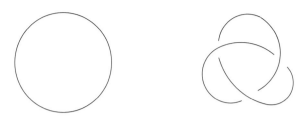

Figure 1.9. *Homeomorphic 1-manifolds.*

1.2. Differentiable Manifolds

Recall the following notion from calculus:

Definition 1.2.1. A map $f : \mathbb{R}^n \to \mathbb{R}^m$ is said to be C^q if it has continuous partial derivatives of order q. A map is said to be *smooth* (or C^∞) if it has partial derivatives of all orders.

Definition 1.2.2. A C^q-*manifold*, for $q \in [0, \infty]$, is a topological manifold M with an atlas that satisfies the additional requirement of being C^q, meaning that for any pair of charts (M_α, ϕ_α), (M_β, ϕ_β) in this atlas, the map $\phi_\beta \circ \phi_\alpha^{-1}$ (where it is defined) is C^q. A C^∞-manifold is also called a *differentiable, or smooth, manifold*.

Given an atlas for a manifold, a map of the form $\phi_\beta \circ \phi_\alpha^{-1}$ is called a *transition map* and is denoted by $\phi_{\alpha\beta}$.

Example 1.2.3. One immediate example of a smooth manifold is again $\mathbb{R}^n$ or open subsets of $\mathbb{R}^n$. These manifolds admit an atlas with a single chart. So the condition on transition maps is vacuously true.

Example 1.2.4. In Example 1.1.4 we saw that $\mathbb{S}^n$ is an n-manifold by exhibiting an atlas with two charts:

$$(\mathbb{S}^n \backslash (0, \ldots, 0, 1), h), \quad (\mathbb{S}^n \backslash (0, \ldots, 0, -1), h')$$

where

$$h(x_1, \ldots, x_{n+1}) = \frac{1}{1 - x_{n+1}}(x_1, \ldots, x_n)$$

and

$$h'(x_1, \ldots, x_{n+1}) = \frac{1}{1 + x_{n+1}}(x_1, \ldots, x_n).$$

Thus to show that $\mathbb{S}^n$ is a smooth manifold, we need to show that the transition maps $h' \circ h^{-1}$ and $h \circ (h')^{-1}$ are smooth. We will only check that

$h' \circ h^{-1}$ is smooth. (The case $h \circ (h')^{-1}$ is analogous.) Here $h^{-1}(y_1, \ldots, y_n) =$
$$\left(\frac{2y_1}{1+y_1^2+\cdots+y_n^2}, \ldots, \frac{2y_n}{1+y_1^2+\cdots+y_n^2}, \frac{-1+y_1^2+\cdots+y_n^2}{1+y_1^2+\cdots+y_n^2} \right),$$
$$h' \circ h^{-1}(y_1, \ldots, y_n) = \frac{1}{y_1^2+\cdots+y_n^2}(y_1, \ldots, y_n).$$

It follows that $h' \circ h^{-1}$ is smooth except at the origin where the composition of maps is not defined. Thus $\mathbb{S}^n$ is a smooth manifold.

Example 1.2.5. In the exercises you proved that the product of manifolds is a manifold. Since the product of smooth maps is smooth, the product of smooth manifolds is a smooth manifold. It follows that $\mathbb{T}^n$ is a smooth manifold.

In calculus we learn about differentiable maps from $\mathbb{R}^n$ to $\mathbb{R}^m$. Some concepts extend to manifolds.

Definition 1.2.6. Let M be a manifold with atlas $\{(M_\alpha, \phi_\alpha)\}$ and let N be a manifold with atlas $\{(N_\beta, \psi_\beta)\}$. We say that the map $f : M \to N$ is C^q if $\forall \alpha, \beta$, the map $\psi_\beta \circ f \circ \phi_\alpha^{-1}$ (where it is defined) is C^q.

Definition 1.2.7. A C^q-map between C^q-manifolds with a C^q-inverse is called a C^q-*diffeomorphism*. A C^∞-diffeomorphism is simply called a *diffeomorphism*.

Remark 1.2.8. The map $f : \mathbb{R} \to \mathbb{R}$ given by $f(x) = x^3$ is a C^∞-map but is not a diffeomorphism because its derivative is singular at 0. (In fact, it is not even a C^1-diffeomorphism.)

Definition 1.2.9. Two C^q-manifolds are considered *equivalent* if there is a C^q-diffeomorphism between them.

In the exercises, you will extend the notion of submanifold, manifold with boundary, and submanifold of a manifold with boundary to the DIFF category, that is, to the setting in which manifolds are considered in this section.

In the DIFF category we are interested in smooth maps between manifolds.

Example 1.2.10. Projection from $\mathbb{T}^2 = \mathbb{S}^1 \times \mathbb{S}^1$ onto the second factor is a smooth map between manifolds.

In Appendix A, we introduce the notion of *transversality* in the category of DIFF manifolds. Another concept that is best described in this category is that of a Morse function. We discuss the concept in more detail in Appendix B but provide the basic definition here.

Definition 1.2.11. Let M be a C^q-manifold for $q \geq 1$, $x \in M$, and (M_α, ϕ_α) a chart with $x \in M_\alpha$. We say that x is a critical point of a function $f : M \to \mathbb{R}$ if it is a critical point of $f \circ \phi_\alpha^{-1}$.

Definition 1.2.12. A critical point of a function $g : \mathbb{R}^n \to \mathbb{R}$ is *non-degenerate* if the Hessian of g is non-singular at x. For $M, x, (M_\alpha, \phi_\alpha), f$ as above, we say that x is a *non-degenerate* critical point of f if it is a non-degenerate critical point of $f \circ \phi_\alpha^{-1}$.

Remark 1.2.13. In the exercises, you will verify that these two definitions do not depend on the chart used.

Definition 1.2.14. A *Morse function* on a manifold M is a smooth function $f : M \to \mathbb{R}$ that satisfies the following:

- f has only non-degenerate critical points;
- distinct critical points of f take on distinct values.

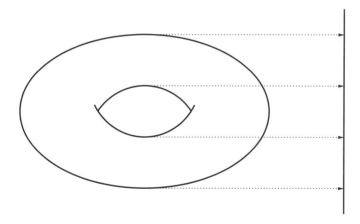

Figure 1.10. *Projection onto the z-axis.*

Example 1.2.15. The torus pictured in Figure 1.10 sits in $\mathbb{R}^3$ and projection onto the third coordinate defines a function. This function has four critical points: (1) a minimum; (2) two saddle points; (3) a maximum. You will verify in the exercises that all these critical points are non-degenerate. Since they occur at distinct levels, this function is a Morse function.

Exercises

Exercise 1. Verify that the definitions of critical point and non-degenerate critical point for a function $f : M \to \mathbb{R}$ are independent of the chart used.

Exercise 2. Verify that the critical points of the function given in Example 1.2.15 are non-degenerate.

Exercise 3. Find other Morse functions on the torus by drawing other pictures representing the torus in $\mathbb{R}^3$.

Exercise 4. Compute the quantity #minima+#maxima−#saddles for the Morse function in Example 1.2.15 and the Morse functions from Exercise 3.

Exercise 5. Extend the notions of submanifold, manifold with boundary, and submanifold of a manifold with boundary to the DIFF category.

1.3. Oriented Manifolds

The notion of orientation allows us to partition manifolds into two types: orientable and non-orientable. We introduce the notion here in the DIFF category, though analogous notions exist in some other categories, most notably TOP (discussed above) and TRIANG (discussed below).

Definition 1.3.1. A C^∞-manifold M with boundary is *orientable* if it has an atlas such that the Jacobians of all transition maps have positive determinant. Otherwise M is *non-orientable*. An *orientation* of M is such an atlas. We often write $(M, \{\phi_\alpha\})$ to denote an oriented manifold.

Example 1.3.2. The annulus $\mathbb{A} = \mathbb{S}^1 \times (-1, 1)$ is orientable. It can be covered by two charts:

$$\left\{ \left(\left\{ (e^{ix}, t) : -\frac{\pi}{4} < x < \frac{5\pi}{4}, -1 < t < 1 \right\}, \phi_1 \right), \right.$$
$$\left. \left(\left\{ (e^{ix}, t) : \frac{3\pi}{4} < x < \frac{9\pi}{4}, -1 < t < 1 \right\}, \phi_2 \right) \right\}$$

where $\phi_i((e^{ix}, t)) = (x, t)$. The transition maps are

$$\phi_2 \circ \phi_1^{-1}((x, t)) = \begin{cases} (x + 2\pi, t) & \text{if } -\frac{\pi}{4} < x < \frac{\pi}{4}, \\ (x, t) & \text{if } \frac{3\pi}{4} < x < \frac{5\pi}{4} \end{cases}$$

and

$$\phi_1 \circ \phi_2^{-1}((x, t)) = \begin{cases} (x - 2\pi, t) & \text{if } \frac{7\pi}{4} < x < \frac{9\pi}{4}, \\ (x, t) & \text{if } \frac{3\pi}{4} < x < \frac{5\pi}{4}. \end{cases}$$

Thus the Jacobians of all transition maps (where they are defined) have positive determinant.

Example 1.3.3. The Möbius band is the surface with boundary obtained by identifying two sides of a rectangle, as pictured in Figure 1.11. It is non-orientable. Intuitively speaking, this is because in an orientable surface, there is a well-defined notion of, for instance, "above", or "being to the

1.3. Oriented Manifolds

right". (You can make sense out of this in $\mathbb{R}^2$. The condition on Jacobians guarantees that the notion patches together correctly on overlapping charts.) Now consider the core curve of the Möbius band, as pictured in Figure 1.12, and imagine a curve "above" or "to the right" of the core curve. Such a curve does not exist, and so the Möbius band is non-orientable.

Figure 1.11. *The Möbius band.*

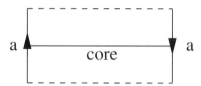

Figure 1.12. *The core of the Möbius band.*

Let M be a differentiable manifold with two orientations $(M, \{\phi_\alpha\})$ and $(M, \{\psi_\beta\})$. The subset of M on which $\phi_\alpha \circ \psi_\beta^{-1}$ is defined and has a Jacobian with a positive determinant is the subset where the orientations of $(M, \{\phi_\alpha\})$ and $(M, \{\psi_\beta\})$ are said to *coincide*. The subset where $\phi_\alpha \circ \psi_\beta^{-1}$ is defined and has a Jacobian with a negative determinant is the subset where the orientations of $(M, \{\phi_\alpha\})$ and $(M, \{\psi_\beta\})$ are said to *differ*. These two sets are open. Thus for a connected manifold M, two orientations either coincide on all of M or differ on all of M.

Given an oriented manifold $(M, \{\phi_\alpha\})$, we can create an oriented manifold $(M, \{\psi_\alpha\})$ for which the orientations differ on all of M. Indeed, for each $\phi_\alpha : U_\alpha \to \mathbb{R}^n$, defined by $\phi_\alpha(x) = (x_1, x_2, \ldots, x_n)$, we substitute $\psi_\alpha : U_\alpha \to \mathbb{R}^n$, defined by $\psi_\alpha(x) = (-x_1, x_2, \ldots, x_n)$. The resulting orientation is called the *opposite* orientation of the original one. We denote the resulting oriented manifold by $-M$.

If M is an oriented manifold with boundary, then there is a natural orientation on ∂M, called the *induced orientation of ∂M*. See for instance [48]. We will not discuss the technicalities here.

On a related note, given an oriented manifold M, we sometimes assign an orientation to an orientable submanifold. When the dimension of the submanifold is one less than the dimension of the manifold in which it lies, then choosing an orientation on the submanifold is equivalent to choosing a smoothly varying normal direction. For details see [48].

Definition 1.3.4. For oriented C^∞-manifolds $(M, \{\phi_\alpha\}), (N, \{\psi_\beta\})$ of the same dimension, a smooth map $h : M \to N$ is *orientation-preserving* if the Jacobians of the maps $\psi_\beta \circ h \circ \phi_\alpha^{-1}$ (where they are defined) all have positive determinant. If they all have negative determinant it is *orientation-reversing*.

Example 1.3.5. The map $f : \mathbb{S}^1 \to \mathbb{S}^1$ given by $f(e^{2\pi i x}) = e^{4\pi i x}$ is orientation-preserving.

Example 1.3.6. The map $f : \mathbb{S}^1 \to \mathbb{S}^1$ given by $f(e^{2\pi i x}) = e^{-2\pi i x}$ is orientation-reversing.

In a non-orientable manifold M each chart defines a local orientation. If c is a closed 1-dimensional submanifold of M, then the transition maps for those charts of M that meet c may or may not have Jacobians with positive determinants. If there is an atlas for M for which all charts that meet c have transition maps whose Jacobians have positive determinants, then we say that c is an *orientation-preserving* closed 1-dimensional submanifold of M. If there is no such atlas, then c is an *orientation-reversing* closed 1-dimensional submanifold of M.

In the exercises, you will prove the following:

Lemma 1.3.7. *The manifold M is non-orientable if and only if M contains an orientation-reversing closed 1-dimensional submanifold.*

Exercises

Exercise 1. Show that $\mathbb{S}^n$ is an orientable manifold.

Exercise 2. Show that the product of two orientable manifolds is an orientable manifold.

Exercise 3. Show that $\mathbb{T}^n$ is an orientable manifold.

Exercise 4. Prove Lemma 1.3.7.

1.4. Triangulated Manifolds

In this section we consider manifolds in the TRIANG category. The definitions that lay the foundation for this study (simplicial complexes and related notions) are discussed in greater detail in [148]. See also [102].

Note that in Definitions 1.4.12 and 1.4.13 we make a subtle departure from the traditional definitions. This is in line with a contemporary view of

1.4. Triangulated Manifolds

triangulations. For more detail, see Chapter 5, where we will reconcile the two notions.

Definition 1.4.1. Denote the $(k+1)$-tuple in $\mathbb{R}^{k+1}$ with i-th entry 1 and all other entries 0 by v_i. The set

$$\left\{ a_0 v_0 + \cdots + a_k v_k \; : \; a_0 \geq 0, \ldots, a_k \geq 0, \sum_{i=0}^{k} a_i = 1 \right\}$$

is called the *standard (closed) k-simplex* and is denoted by $[v_0, \ldots, v_k]$ or simply by $[s]$. The *dimension* of the standard k-simplex is k.

The set

$$\left\{ a_0 v_0 + \cdots + a_k v_k \; : \; a_0 > 0, \ldots, a_k > 0, \sum_{i=0}^{k} a_i = 1 \right\}$$

is called the *standard open k-simplex* and is denoted by $(v_0, \ldots, v_k)$ or simply by (s). We also call (s) the *interior* of $[s]$.

Example 1.4.2. The standard 0-simplex is the point $1 \in \mathbb{R}$. This is also the standard open 0-simplex.

Definition 1.4.3. A k-*simplex* in a topological space X is a continuous map $f : [s] \to X$ such that $[s]$ is the standard k-simplex and the restriction $f|_{(s)}$ is a homeomorphism onto its image. An *open k-simplex* is the restriction of a k-simplex to the interior (s) of $[s]$.

Abusing notation slightly, we will often refer to the image of f as a k-simplex.

Example 1.4.4. The standard 1-simplex is homeomorphic to an interval. See Figures 1.13 and 1.14.

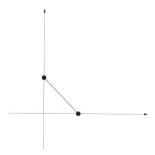

Figure 1.13. *The standard 1-simplex.*

Figure 1.14. *A 1-simplex.*

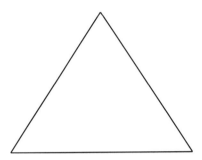

Figure 1.15. *A 2-simplex.*

Example 1.4.5. The standard 2-simplex is a "triangle" with vertices at $(1,0,0)$, $(0,1,0)$, and $(0,0,1)$. Figure 1.15 depicts a 2-simplex.

Example 1.4.6. The standard 3-simplex is a tetrahedron with vertices at $(1,0,0,0)$, $(0,1,0,0)$, $(0,0,1,0)$, and $(0,0,0,1)$. Figure 1.16 depicts a 3-simplex.

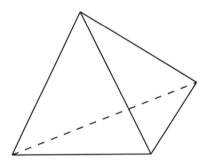

Figure 1.16. *A 3-simplex.*

Definition 1.4.7 (Faces). For $j = 0, \ldots, k$, a *face* of the standard k-simplex $[s]$ is a subset of $[s]$ of the form
$$\{a_0 v_0 + \cdots + a_k v_k : a_{i_1} = 0, \ldots, a_{i_j} = 0\}.$$
The *dimension* of the face is $k - j$. The faces of the k-simplex $f : [s] \to X$ are the restriction maps $f|_{[t]}$ for $[t]$ a face of the standard k-simplex $[s]$. A 0-dimensional face of a simplex is also called a *vertex*. A 1-dimensional face of a simplex is also called an *edge*.

1.4. Triangulated Manifolds

Example 1.4.8. A 0-simplex has only itself as a face.

Example 1.4.9. The standard 1-simplex $[v_0, v_1]$ (and hence all 1-simplices) has itself as a 1-dimensional face. It also has two 0-dimensional faces, $[v_0]$ and $[v_1]$.

Example 1.4.10. A 2-simplex has one 2-dimensional, three 1-dimensional, and three 0-dimensional faces.

Example 1.4.11. A 3-simplex has one 3-dimensional, four 2-dimensional, six 1-dimensional, and four 0-dimensional faces.

Definition 1.4.12 (Simplicial complex). A *simplicial complex* based on the topological space X is a set of simplices
$$K = \{f : [s] \to X\}$$
in the topological space X such that

(1) $\forall$ simplices $f \in K$, all faces of f are in K;

(2) $\forall$ simplices $f_1, f_2 \in K$, $\text{im}(f_1|_{(s_1)}) \cap \text{im}(f_2|_{(s_2)}) \neq \emptyset \implies \text{im}(f_1|_{(s_1)}) = \text{im}(f_2|_{(s_2)})$.

The *dimension* of a simplicial complex K is the supremum of the dimensions of the simplices in K. We denote the union of the images of the simplices in K by $|K|$ and call $|K|$ the *underlying space* of K.

For examples see Figures 1.17, 1.18, and 1.19. Traditionally, one also requires that each closed simplex be embedded and that two distinct closed simplices meet in at most one face. We do not make these assumptions here!

Figure 1.17. *A 1-dimensional simplicial complex.*

Definition 1.4.13. A *triangulated n-manifold* is a pair (M, K), where M is a topological n-manifold and K is a simplicial complex based on M such that

- $|K| = M$;
- K is locally finite, i.e., for every compact subset C of M, the set $\{f \in K : C \cap \text{im } f \neq \emptyset\}$ is finite;

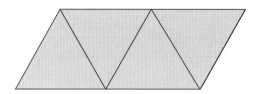

Figure 1.18. *A 2-dimensional simplicial complex.*

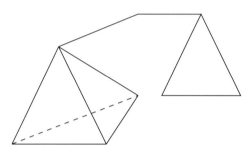

Figure 1.19. *A 3-dimensional simplicial complex.*

- for $f, g \in K$, restricted to open simplices, the map $g^{-1} \circ f$ is affine on its domain.

Here K is called the *triangulation* of M. For $f : [s] \to M$ an n-simplex in K, the pair $(\operatorname{im} f, f^{-1})$ is called a *simplicial chart* of the triangulation. In this context, we often write $K = \{f_\alpha\}$ and denote the collection of all charts on M by $\{(\operatorname{im} f_\alpha, f_\alpha^{-1})\}$. We also refer to f_α as a *simplex* of M.

There are related categories of manifolds (piecewise linear (PL), combinatorial) that we will not discuss here. Suffice it to say that some distinctions are subtle.

Example 1.4.14. Figure 1.20 depicts a triangulation of $\mathbb{S}^2$.

Example 1.4.15. Figure 1.21 depicts a triangulation of $\mathbb{T}^2$ with two 2-simplices.

Example 1.4.16. Figure 1.22 depicts a triangulated cube. It can be interpreted as a portion of a triangulation of $\mathbb{T}^3$. We obtain a triangulation of $\mathbb{T}^3$ by identifying eight appropriately chosen reflections of this cube. The resulting triangulation contains forty 3-simplices.

Theorem 1.4.17. *Every compact 1-manifold admits a triangulation.*

1.4. Triangulated Manifolds

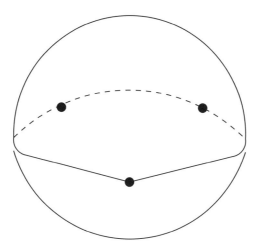

Figure 1.20. *A triangulation of $\mathbb{S}^2$.*

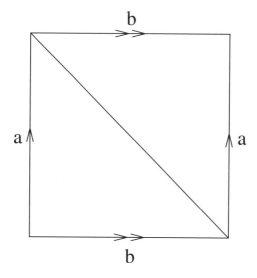

Figure 1.21. *A triangulation of $\mathbb{T}^2$.*

Proving the above theorem is an easy exercise in understanding compact 1-manifolds. Analogous theorems for 2- and 3-manifolds also hold but are much harder to prove.

Theorem 1.4.18 (T. Radó, B. Kerekjarto)**.** *Every compact 2-manifold admits a triangulation.*

A proof can be found in [**4**].

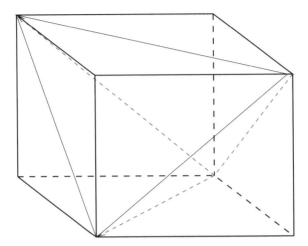

Figure 1.22. *A portion of a triangulation of* $\mathbb{T}^3$.

Theorem 1.4.19 (R. H. Bing, E. Moise). *Every compact 3-manifold admits a triangulation.*

A proof can be found in [**101**].

This theorem has been put to extensive use in the study of 3-manifolds. In view of Lemma 1.4.28 below, it allows us to build 3-manifolds by identifying 3-simplices along faces. In Chapter 5 we will see a brief introduction to normal surface theory (after Wolfgang Haken) where this point of view bears fruit.

Definition 1.4.20. Let K and L be simplicial complexes. We say that a continuous map $\phi : |K| \to |L|$ is a *simplicial map* if for every simplex f in K, there is a simplex g in L such that $\phi \circ f = g$.

Informally speaking, a simplicial map maps simplices to simplices, but the latter might be of lower dimension than the former.

Example 1.4.21. Figure 1.23 depicts a simplicial map from the top simplicial complex (containing four triangles) to the bottom simplicial complex (containing two edges): The two vertices on the left go to the left vertex below, the two vertices in the middle go to the middle vertex below, and the two vertices on the right go to the right vertex below. Extend this map of vertices linearly over the 2-simplices to obtain a simplicial map.

Non-Example. A map that is not simplicial is depicted in Figure 1.24. It maps the top simplicial complex (containing four triangles) to the bottom simplicial complex (containing one edge) via nearest point projection (in $\mathbb{R}^2$).

1.4. Triangulated Manifolds

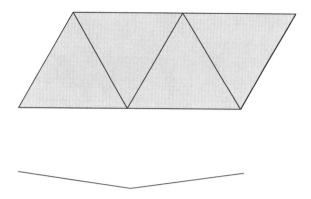

Figure 1.23. *A simplicial map.*

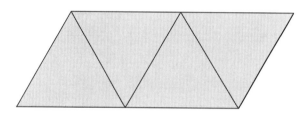

Figure 1.24. *A map between simplicial complexes that is not simplicial.*

Definition 1.4.22. Let K_1, K_2 be simplicial complexes and let $\phi : |K_1| \to |K_2|$ be a map. We say that ϕ is a *simplicial isomorphism* if it is simplicial and a homeomorphism. Two simplicial complexes K_1, K_2 are *isomorphic* if there is a simplicial isomorphism $\phi : |K_1| \to |K_2|$.

Definition 1.4.23. Let (M_1, K_1) and (M_2, K_2) be triangulated n-manifolds. The two triangulated n-manifolds are considered *equivalent* if there is a simplicial isomorphism $\phi : M_1 \to M_2$.

Definition 1.4.24. A *subcomplex* of a simplicial complex K is a simplicial complex L such that $f \in L$ implies $f \in K$. We write $L \subset K$.

Example 1.4.25. The unshaded area in Figure 1.25 depicts a subcomplex of the simplicial complex depicted in Figure 1.18.

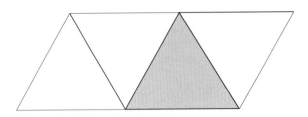

Figure 1.25. *A subcomplex of a simplicial complex.*

One concept that is easily defined in the TRIANG category is the following:

Definition 1.4.26. The *Euler characteristic* of a finite simplicial complex K of dimension k is computed via the following formula:

$$\chi(K) = \sum_{i=0}^{k}(-1)^i \#\{\text{simplices of dimension } i \text{ in } K\}.$$

Remark 1.4.27. In fact, Definition 1.4.26 is well-defined for the underlying space of a simplicial complex. This can be established via a computation involving the notion of homology from algebraic topology.

The standard k-simplex lies in $\mathbb{R}^{k+1}$ and is thus a topological space. Suppose that (M, K) is a triangulated k-manifold and consider a collection $\mathcal{C}$ that is the disjoint union of copies of standard k-simplices, one copy for each k-simplex f in K. Now construct a quotient space from $\mathcal{C}$ as follows: We identify points in the collection $\mathcal{C}$ if and only if the corresponding points $x \in [s]$ and $y \in [t]$ for $f : [s] \to M$ and $g : [t] \to M$ in K satisfy $f(x) = g(y)$. We leave the proof of the following lemma as an exercise.

Lemma 1.4.28. *The quotient space M' (in the quotient topology) is homeomorphic to the topological manifold M.*

The observation in Lemma 1.4.28 allows us to think of triangulated n-manifolds as objects that are constructed out of standard simplices by identifying faces.

Example 1.4.29. Figure 1.26 describes a triangulation of $\mathbb{S}^3$. The labeling gives the identification of the edges. 2-dimensional faces are identified if all three of their edges are identified.

1.5. Geometric Manifolds

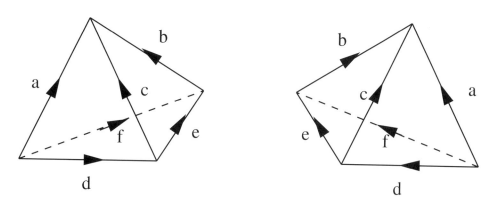

Figure 1.26. *A triangulation of* $\mathbb{S}^3$.

Exercises

Exercise 1. Prove that a simplicial isomorphism has an inverse that is also a simplicial isomorphism.

Exercise 2. Prove Lemma 1.4.28. (Hint: The simplices in M piece together to give a continuous bijection $h : M \to M'$. To prove that h^{-1} is also continuous, use local finiteness of K to establish that h^{-1} is a closed map.)

Exercise 3. Find a triangulation of the projective plane.

Exercise 4. How many triangulations of the 2-torus can you find (e.g., 0, 1, ∞)?

Exercise 5. Prove that every compact 1-manifold admits a triangulation.

Exercise 6. Consider the triangulation of $\mathbb{S}^2$ in Figure 1.20. Try to add vertices and edges to create a triangulation such that any pair of 2-dimensional simplices meets in at most one edge.

Exercise 7. Compute the Euler characteristic of the sphere and of the torus by choosing triangulations.

1.5. Geometric Manifolds

Geometry has evolved since Euclid's time. Euclidean geometry and the other structures now also considered geometries provide insight into manifolds. We will assume a basic familiarity with Euclidean geometry in this section. In Chapter 7, we will develop the basics of hyperbolic geometry. While in this section we focus on Euclidean manifolds, the definitions carry over verbatim with the term "Euclidean" replaced by the term "hyperbolic" in order to define hyperbolic manifolds. For more on geometric 3-manifolds, see [**145**], [**154**], [**153**], or [**126**].

Definition 1.5.1. An *isometry* is an invertible map between metric spaces that preserves distances. A Euclidean isometry, i.e., an isometry $f : \mathbb{R}^n \to \mathbb{R}^n$, is also called a *rigid motion*.

Definition 1.5.2. A *Euclidean manifold* is a topological manifold M with an atlas that satisfies the additional requirement that the transition maps are restrictions of Euclidean isometries. We often write $(M, \{\phi_\alpha\})$ to denote a Euclidean manifold.

The more general term is that of a *geometric manifold*. There the additional requirement is that the transition maps are isometries in a given geometry.

Example 1.5.3. An immediate example of a Euclidean manifold is again $\mathbb{R}^n$. This manifold admits an atlas with a single chart. So the condition on transition maps is vacuously true.

Example 1.5.4. It follows from the description of $\mathbb{T}^n$ given in Example 1.1.5 that $\mathbb{T}^n$ is a Euclidean manifold.

Definition 1.5.5. Given two Euclidean manifolds $(M, \phi_\alpha), (N, \phi_\beta)$, a map $f : M \to N$ is an *isometry* if $\forall \alpha, \beta$ the maps $\phi_\alpha^{-1} \circ f \circ \psi_\beta$ are restrictions of Euclidean isometries.

Definition 1.5.6. Two Euclidean manifolds M, N are considered *equivalent* if there is an invertible isometry $h : M \to N$.

Remark 1.5.7. On a connected Euclidean manifold there is a well-defined metric. Metric concepts such as completeness, diameter, volume, etc., are thus defined for Euclidean manifolds.

Exercises

Exercise 1. Show that among Euclidean 2-manifolds there are infinitely many inequivalent 2-tori.

Exercise 2. The definition of Euclidean manifolds extends to manifolds with boundary. Show that among Euclidean manifolds with boundary there are infinitely many inequivalent annuli.

Exercise 3. Show that $\mathbb{S}^1$ is a Euclidean manifold.

Exercise 4*. Show that $\mathbb{S}^n$ is not a Euclidean manifold for $n > 1$.

1.6. Connected Sums

The following two notions are fundamental to the study of manifolds.

Definition 1.6.1. Two continuous maps $f_0, f_1 : M \to N$ are *homotopic* if there is a continuous map $H : M \times [0,1] \to N$ such that $H(x,0) = f_0$ and $H(x,1) = f_1(x)$ $\forall x \in M$. The map H is called a *homotopy* between f_0 and f_1.

Definition 1.6.2. Two embeddings $f_0, f_1 : M \to N$ are *isotopic* if there is a continuous map $H : M \times [0,1] \to N$ such that $H(x,0) = f_0$ and $H(x,1) = f_1(x)$ $\forall x \in M$ and such that $\forall t \in [0,1]$, the map f_t defined by $H(\ , t)$ is an embedding. The map H is called an *isotopy* between f_0 and f_1.

Two submanifolds S_0, S_1 of M are *isotopic* if their inclusion maps are isotopic.

The following two theorems are due to Gugenheim (see [**47**]) in the TRIANG category and are also a consequence of the Isotopy Uniqueness of Regular Neighborhoods Theorem due to Rourke and Sanderson (see [**128**]). They are fundamental theorems in the study of n-manifolds. We will need these theorems in the discussion of Definition 1.6.5.

Theorem 1.6.3. *Every orientation-preserving homeomorphism of an n-ball or n-sphere is isotopic to the identity.*

We will prove a special case of Theorem 1.6.3, known as the Alexander Trick, in Section 2.5.

Theorem 1.6.4. *If B_1, B_2 are n-balls in the interior of a connected n-manifold M, then there is an isotopy $f : M \times I \to M$ such that $f(\ ,0)|_{B_1}$ is the identity and $f(\ ,1)|_{B_1}$ is a homeomorphism onto B_2.*

The proof of Theorem 1.6.4 is left as an exercise. Theorem 1.6.4 ensures that Definition 1.6.5 is well-defined. Orientations also play a subtle role in Definition 1.6.5.

Definition 1.6.5. Let M_1, M_2 be n-manifolds. Delete small open n-balls B_1 from M_1 and B_2 from M_2. Identify M_1 and M_2 along the resulting $(n-1)$-sphere boundary components. This results in an n-manifold called a *connected sum* of M_1 and M_2 and is denoted by $M_1 \# M_2$. In the case that M_1 and M_2 are oriented, we further require that the identification of the boundaries of B_1 and B_2 be via an orientation-reversing homeomorphism (with respect to the induced boundary orientations on $\partial B_1, \partial B_2$).

The connected sum $M_1 \# M_2$ is *trivial* if either M_1 or M_2 is $\mathbb{S}^n$. Otherwise, $M_1 \# M_2$ is *non-trivial*.

In deleting the small open n-balls, we are making a choice, but Theorem 1.6.4 ensures that this choice is inconsequential. If M_1, M_2 are oriented, then Theorem 1.6.3 tells us that any two choices of identification of ∂B_1 and $-\partial B_2$ are isotopic and it follows that the manifolds obtained via this identification are homeomorphic. Thus for oriented manifolds M_1, M_2, there is a unique connected sum (i.e., connected sum is well-defined in the oriented category).

For orientable (but not oriented) manifolds it is possible to have two non-homeomorphic connected sums of M_1 and M_2. Specifically, endow M_1, M_2 with orientations and consider $M_1 \# M_2$ versus $M_1 \# (-M_2)$. For examples of oriented 3-manifolds M_1, M_2 for which $M_1 \# M_2 \neq M_1 \# (-M_2)$, see [**63**].

If at least one of M_1, M_2, say M_1, is non-orientable, then there is, in fact, a unique connected sum $M_1 \# M_2$. Indeed, this is because every non-orientable n-manifold M contains an orientation-reversing closed 1-dimensional submanifold (by Lemma 1.3.7). Isotoping a small n-ball along such a 1-dimensional submanifold reverses the local orientation on an open neighborhood of the n-ball B_1 and hence also the induced boundary orientation of $(M_1 \backslash B_1) \backslash \partial M_1$. This isotopy extends to an isotopy between the two possibilities for connected sums of M_1 and M_2, thereby demonstrating that they are homeomorphic.

In the case of surfaces, the situation is simpler. Every oriented surface S admits an orientation-reversing homeomorphism $h : S \to S$. Thus for any pair S_1, S_2 of oriented surfaces, $S_1 \# S_2 = S_1 \# h(S_2) = S_1 \# - S_2$. Hence for orientable surfaces, the choice of identifying homeomorphism is inconsequential and the connected sum is unique.

A word on notation: We denote repeated connected sums by $M = M_1 \# \cdots \# M_n$. We leave it as an exercise to show that connected sum of manifolds is associative and commutative. (An example to consider is the following: The connected sum of four 3-manifolds $M_1, \ldots, M_4$ can be obtained in several ways. For instance, take the connected sum by removing three small 3-balls from M_1, one small 3-ball from M_2, M_3, M_4 and identifying the resulting boundary components of M_1 with those of M_2, M_3, M_4. On the other hand, take the connected sum by removing one small 3-ball from M_1 and M_4, two small 3-balls from M_2 and M_3 and identifying the resulting boundary component of M_1 with one of M_2's, the resulting boundary component of M_4 with one of M_3's, and the remaining resulting boundary components of M_2 and M_3 with each other. Figure 1.27 hints at why the repeated connected sum is nevertheless well-defined.)

Definition 1.6.6 (Prime n-manifold). An n-manifold M is *prime* if $M = M_1 \# M_2$ implies that either M_1 or M_2 is the n-sphere, i.e., that the connected sum $M = M_1 \# M_2$ is trivial.

1.7. Equivalence of Categories

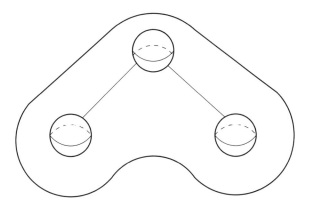

Figure 1.27. *The spheres in a prime decomposition.*

Remark 1.6.7. An n-manifold is prime if and only if it contains no separating essential $(n-1)$-sphere.

Exercises

Exercise 1. Prove Theorem 1.6.4.

Exercise 2. Compute the Euler characteristic of $\mathbb{T}^2 \# \mathbb{R}P^2$.

Exercise 3. Prove that if S_1, S_2 are surfaces, then
$$\chi(S_1 \# S_2) = \chi(S_1) + \chi(S_2) - 2.$$

Exercise 4. Prove that connected sum is associative and commutative.

Exercise 5. Is $\mathbb{S}^2 \times \mathbb{S}^1$ a non-trivial connected sum?

1.7. Equivalence of Categories

We have discussed manifolds from several points of view. The key idea, except in the TRIANG category, is the following: Choose a collection G of invertible maps of $\mathbb{R}^n$ that is closed under group operations and require the transition maps to belong to G. The PL category is similar to the TRIANG category, but there are subtle differences that we do not wish to discuss here.

There are natural maps from all the categories considered here to TOP (in each case, it is the forgetful functor). Amazingly, for 2-manifolds and 3-manifolds, these maps are isomorphisms in the case of DIFF and TRIANG! This means that every topological 3-manifold admits a triangulation.

Furthermore, any two triangulations of a topological 3-manifold admit a common subdivision. Much of the work establishing the equivalence of the categories TOP and TRIANG is due to E. Moise and R. H. Bing. See [**102**] and [**101**].

It deserves to be mentioned that this equivalence of categories is not true for higher-dimensional manifolds. This was first shown by Milnor. He exhibited examples of 7-spheres with distinct differentiable structures, thus demonstrating that DIFF $\neq$ TOP for 7-dimensional manifolds. To date, it is unknown whether there are smooth structures on S^4 that are not diffeomorphic. A candidate for an "exotic" smooth structure on the 4-sphere was given by Scharlemann. See [**132**]. Decades later it was proved by Akbulut that this smooth structure is in fact diffeomorphic to the standard smooth structure on the 4-sphere. See [**5**].

The advantage of the equivalence of the categories TOP, DIFF, and TRIANG in the case of 3-manifolds is that we can introduce a concept or prove a theorem in the category that best suits the concept or theorem. For instance, we introduced the notion of a Morse function in the DIFF category and we introduced the notion of Euler characteristic in the TRIANG category. The notions can be introduced in the other categories as well, but the definitions are necessarily more cumbersome, so we will not do so here. We will be moving back and forth between categories when this is convenient. When categories are equivalent, it is acceptable, though not esthetically pleasing, to prove theorems using concepts introduced in distinct categories.

Recall, for instance, the definition of Euler characteristic in Section 1.4. We will use the notion of Euler characteristic in the other categories as well. For instance, given a smooth surface S, we can compute the Euler characteristic by triangulating the surface and taking the specified alternating sum. If S contains a finite graph Γ such that $S\backslash\Gamma$ consists of a finite number of disks, then the Euler characteristic can be calculated directly from Γ and $S\backslash\Gamma$. You will prove this in the exercises.

We can compute the Euler characteristic not only for closed surfaces but also punctured surfaces. Specifically, since points have Euler characteristic 1, the result of removing n points from a surface S is $\chi(S) - n$. One theorem we will use repeatedly, stated here in the case of 2-manifolds, is the following:

Theorem 1.7.1 (Poincaré-Hopf Index Theorem). *Let S be a surface and let $h : S \to \mathbb{R}$ be a Morse function. Then*

$$\chi(S) = \#(minima\ of\ h) + \#(maxima\ of\ h) - \#(saddles\ of\ h).$$

For a proof of this therem, see [**48**].

Exercises

Exercise 1. Show that, even in the case of closed 2-manifolds, the categories DIFF and GEOM are not equivalent.

Exercise 2. Prove that if a surface S contains a finite graph Γ with v vertices and e edges and $S\backslash\Gamma$ consists of r disks, then
$$\chi(S) = v - e + r.$$

Exercise 3. Draw pictures to convince yourself that Theorem 1.7.1 is true.

Exercise 4. Explore how distinct structures on manifolds interact. For example, read [**126**].

Chapter 2

Surfaces

Topological 2-manifolds, also known as surfaces, are well understood. See [**93**]. We here give a brief overview of what is known. There are additional structures on surfaces that add interest and provide tools for the study of 3-manifolds. See Sections 2.5 and 2.6 of this chapter. For an in-depth treatment of the topics in Sections 2.5 and 2.6, see [**25**] and [**39**].

2.1. A Few Facts about 1-Manifolds

It is not hard to show that the only compact connected 1-manifolds are the circle and the interval. (You will prove this in the exercises.) In this chapter we will often consider 1-dimensional submanifolds of surfaces. In subsequent chapters we will consider 1-dimensional and 2-dimensional submanifolds of 3-manifolds.

Definition 2.1.1. A *simple closed curve* in a surface S is a compact connected 1-dimensional submanifold of S without boundary. A *simple arc* in a surface S is a compact connected 1-dimensional submanifold of S with non-empty boundary.

Remark 2.1.2. The only compact connected 1-manifold without boundary is the circle. So we also think of a simple closed curve in a surface S as an embedding of the circle into S. The only compact connected 1-manifold with non-empty boundary is the interval. So we also think of a simple arc in S as an embedding of the interval into S.

One of the fundamental theorems about simple closed curves in surfaces, which we will not prove but which we state here from a topologist's point of view, is the following:

Theorem 2.1.3 (Jordan Curve Theorem). *If c is a simple closed curve in $\mathbb{R}^2$, then $\mathbb{R}^2\backslash c$ consists of two components: one open annulus and one open disk. If c is a simple closed curve in $\mathbb{S}^2$, then $\mathbb{S}^2\backslash c$ consists of two open disks.*

All 1-manifolds are orientable. (See Exercise 2 below.) In the case of a simple closed curve, we can think of an orientation as a direction in which the circle is traversed. In what follows we will be interested in assigning orientations to points of intersection of 1-dimensional submanifolds in a surface.

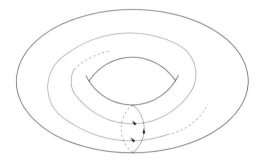

Figure 2.1. *Two points of intersection with the same orientation.*

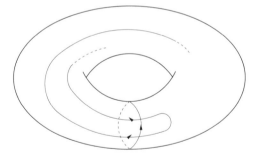

Figure 2.2. *Two oppositely oriented points of intersection.*

Definition 2.1.4. Let a, b be simple closed curves in a surface S. We say that a, b intersect *transversely* and write $a \pitchfork b$ if $\forall x \in a \cap b$, there is a chart (U_γ, ϕ_γ) with $\phi_\gamma(U_\gamma) = \mathbb{R}^2$ such that ϕ_γ maps x to $(0,0)$, a to the x-axis, and b to the y-axis.

It is not hard to see that in an oriented surface S, the chart (U_γ, ϕ_γ) can be chosen to be in the orientation of S. (Postcompose with a reflection in $\mathbb{R}^2$.) It is harder to see that for any pair (a, b) of simple closed curves in a surface S, a can be isotoped to a simple closed curve a' so that $a' \pitchfork b$.

Indeed, this is a consequence of Theorem A.0.19. For the more general definition of transversality, see Appendix A.

Definition 2.1.5. Let a, b be oriented 1-dimensional submanifolds in the oriented surface S such that $a \pitchfork b$. An *orientation* for a point $x \in a \cap b$ is the assignment ± 1 obtained as follows: Let (U_γ, ϕ_γ) be a chart in the orientation of S with $\phi_\gamma(U_\gamma) = \mathbb{R}^2$ such that ϕ_γ maps x to $(0,0)$, a to the x-axis, and b to the y-axis. If $\phi_\gamma|_a$ and $\phi_\gamma|_b$ are both orientation-preserving or both orientation-reversing, then the orientation on x is $+1$. Otherwise, it is -1. See Figures 2.1 and 2.2.

Definition 2.1.6. Let a, b be oriented simple closed curves in the oriented surface S such that $a \pitchfork b$. The sum of the orientations on points in $a \cap b$ is called the *oriented intersection number* of a and b.

A straightforward but technical result that can be found for instance in [48] is the following:

Theorem 2.1.7. *Suppose that a, b, a', b' are oriented simple closed curves in an oriented surface S. Suppose further that a is homotopic to a', b is homotopic to b', $a \pitchfork b$, and $a' \pitchfork b'$. Then the oriented intersection number of a and b is equal to the oriented intersection number of a' and b'.*

Definition 2.1.8. For homotopy classes $[a], [b]$, the *oriented intersection number*, $[a] \cdot [b]$, is the oriented intersection number of transverse representatives of the classes.

In particular, for a homotopy class $[c]$, the oriented intersection number of $[c]$ with itself is the oriented intersection number of representatives c, c' of $[c]$ that intersect transversely.

Example 2.1.9. In $\mathbb{T}^2$, the simple closed curve $m = \mathbb{S}^1 \times \{0\}$ is homotopic to a disjoint simple closed curve m'. Thus $[m] \cdot [m]$ is 0.

Exercises

Exercise 1. Show that the only compact connected 1-manifolds are the circle and the interval.

Exercise 2. Prove that all 1-manifolds are orientable.

Exercise 3. Generalize Example 2.1.9 to arbitrary oriented surfaces.

2.2. Classification of Surfaces

In Chapter 1 we discussed several examples of surfaces, among them the sphere, the torus, the projective plane, the disk, and the annulus. The sphere and the torus are orientable, whereas the projective plane is not. Two more examples of surfaces deserve to be pointed out: (1) The *Klein bottle*. The Klein bottle can be obtained by identifying the sides of a square as in Figure 2.3. We will denote the Klein bottle by $\mathbb{K}^2$. The Klein bottle contains a Möbius band (see Figure 2.4) and is hence non-orientable. (2) The connected sum of g tori is called the *genus g surface*.

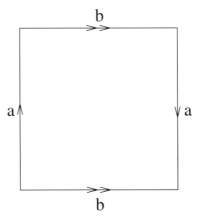

Figure 2.3. *The Klein bottle.*

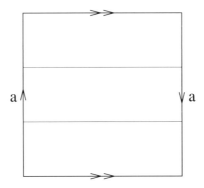

Figure 2.4. *A Möbius band in the Klein bottle.*

Definition 2.2.1. Let S be a connected surface. A simple closed curve c in S is *separating* if $S\backslash c$ has two components. Otherwise it is *non-separating*. A simple closed curve c in a surface S is *inessential* if c is separating and a component of $S\backslash c$ is a disk or annulus. The curve c is *essential* if it is not inessential.

A curve c is non-separating if and only if its complement is connected. Since connectedness is equivalent to path connectedness for manifolds, it

2.2. Classification of Surfaces

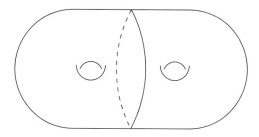

Figure 2.5. *The genus 2 surface as a connected sum of two 2-tori.*

follows that c is non-separating if and only if there is some other simple closed curve that intersects c exactly once.

Lemma 2.2.2. *The closed surface S is prime if and only if S contains no essential separating simple closed curve.*

Proof. If $S = S_1 \# S_2$ is a non-trivial connected sum, then it contains a copy of the curve along which $S_1 \backslash (\text{disk}), S_2 \backslash (\text{disk})$ were identified. See the curve pictured in Figure 2.5. Conversely, if a simple closed curve c in S is essential and separating, then neither component of $S \backslash c$ is a disk. Capping off (i.e., taking the union of) these two components with closed disks creates $S_1 \neq \mathbb{S}^2, S_2 \neq \mathbb{S}^2$ such that $S = S_1 \# S_2$. □

Example 2.2.3. The sphere is a prime surface. This is a consequence of the Jordan Curve Theorem, which tells us that an essential curve as in Lemma 2.2.2 can't exist.

The Klein bottle is not a prime surface. The curve pictured in Figure 2.6 cuts the Klein bottle into two open Möbius bands. Attaching disks to the remnants of the curve in these Möbius bands produces two projective planes. Hence the Klein bottle is the connected sum of two projective planes. The annulus is not a prime surface. It is the connected sum of two disks. More generally, a compact surface with boundary is not prime. It is the connected sum of a closed surface and a finite number of disks.

Triangulations have proved useful in the study of surfaces. Most importantly, they allow us to consider a compact surface as the identification space of a finite number of triangles; see Lemma 1.4.28. For many applications, it is expedient to consider not only triangles but other polygons. In this context, a *polygon* is a convex disk in $\mathbb{R}^2$ whose boundary is partitioned into edges. Specifically, it is often preferable to first identify the triangles to a single polygon. The surface is obtained from this *polygonal representation* by identifying appropriate pairs of edges. In Chapter 1 we encountered

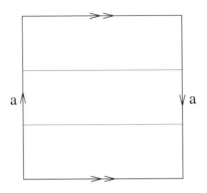

Figure 2.6. $\mathbb{K}^2 = \mathbb{R}P^2 \# \mathbb{R}P^2$.

polygonal representations of the torus and the projective plane. Above, we encountered a polygonal representation of the Klein bottle.

Lemma 2.2.4. *Every compact surface admits a polygonal representation.*

Proof. Recall from Chapter 1 that every surface S admits a triangulation (S, K). The *dual graph* of K is the graph obtained by associating a vertex to each 2-simplex of K and an edge to each 1-simplex of K such that the vertices incident to this edge are the vertex or vertices associated with the 2-simplex or 2-simplices in which the 1-simplex lies.

A *spanning tree* of a graph is a tree that contains all vertices of the graph. Let T be a spanning tree of the dual graph of K. (It exists because K and hence also its dual graph are finite.) In the exercises, you will use induction on the number of vertices in T to construct a polygon P (in $\mathbb{R}^2$) from a finite collection of standard 2-simplices (in $\mathbb{R}^3$), corresponding to the 2-simplices of K, such that the result of identifying appropriate edges of P is the surface S. This provides a polygonal representation of S. □

Lemma 2.2.5. *The sphere, the projective plane, and the torus are prime surfaces.*

Proof.

Case 1. The sphere is prime. See Example 2.2.3.

Case 2. The projective plane.

We argue by contradiction. If $\mathbb{R}P^2$ is not prime, then it contains a separating essential simple closed curve c guaranteed by Lemma 2.2.2. We analyze this curve in the polygonal representation of $\mathbb{R}P^2$ given in Figure 2.7. General position, a consequence of Theorem A.0.19, allows us to assume that c misses the vertices of the polygonal representation and meets the edges transversely.

2.2. Classification of Surfaces

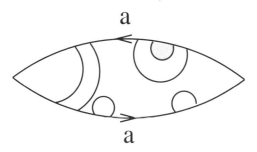

Figure 2.7. *A simple closed curve in $\mathbb{R}P^2$.*

If a subarc, a, of c in this polygonal representation of $\mathbb{R}P^2$ cobounds a disk D together with a subarc, b, of an edge of the boundary of the polygon, then we can assume that a is an outermost such arc, i.e., that the disk D cobounded by a and b has interior disjoint from c. The disk D describes an isotopy that moves a across D to coincide with b and then off of b just beyond b. A disk cut off by an outermost arc is shaded in Figure 2.7. The effect of the isotopy is pictured in Figure 2.8.

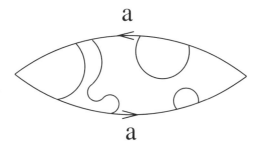

Figure 2.8. *The same simple closed curve in $\mathbb{R}P^2$.*

Proceeding in this manner, we can remove all subarcs of c cobounding disks with subarcs of edges of the polygon. After this isotopy, c consists entirely of subarcs that connect distinct edges of our polygonal representation of $\mathbb{R}P^2$. If c consists of only one such arc, then c is non-separating (see Figure 2.10 where the dashed arc represents a simple closed curve intersecting c once), so this is impossible.

If c consists of more than one arc, consider the two outermost such arcs as in Figure 2.9. The way in which the edges of our polygonal representative of $\mathbb{R}P^2$ are identified forces the endpoints of these two arcs to be identified pairwise. Thus c consists of exactly these two arcs. In this case c cuts $\mathbb{R}P^2$ into a Möbius band and a disk and is hence not essential, so this is also impossible. Hence there is no essential simple closed curve, whence the projective plane is prime.

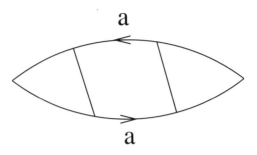

Figure 2.9. *A simple closed curve in $\mathbb{R}P^2$.*

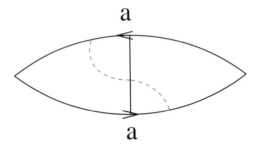

Figure 2.10. *A non-separating simple closed curve in $\mathbb{R}P^2$.*

Case 3. The torus is prime. We leave this case as an exercise.

□

Lemma 2.2.6. *The sphere, the torus, and the projective plane are the only closed prime surfaces.*

Proof. Given a surface S, we consider a polygonal representation of S with polygon P. It may happen that P has two adjacent edges as pictured in Figure 2.11. In this case we modify P by identifying the two edges to obtain a polygon with fewer edges. We continue to denote this polygon by P.

It may also happen that P has two adjacent edges as pictured in Figure 2.12. If there are no other edges, then S is a projective plane. If there are other edges, then the curve represented by the dotted arc in Figure 2.12 must cut off a disk to the other side. (Otherwise S would not be prime.) Hence S is again $\mathbb{R}P^2$.

We may assume, therefore, in the following, that pairs of edges of P that are identified are not adjacent. Let c be a simple closed curve in S that is realized by an arc in P connecting a pair of edges that are identified. (In particular, c intersects no other remnants of edges of P in S.) Suppose first that c is separating. Since S is prime, one of these components, say S_2, must

2.2. Classification of Surfaces

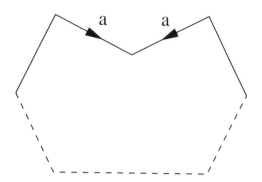

Figure 2.11. *Redundant edges in a polygonal representation of S.*

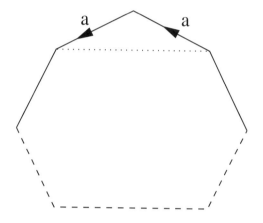

Figure 2.12. *A polygonal representation of S containing a projective plane.*

be a disk. This disk meets remnants of edges of P. This implies that there is a polygonal representation of S with fewer edges.

Now suppose that c is non-separating. Let a be a simple closed curve that intersects c once; see Figure 2.13. Furthermore, let $C(c \cup a)$ be a collar neighborhood (see Appendix A and Figure 2.13) of $c \cup a$.

We leave it as an exercise to show that $C(c \cup a)$ is one of three things: a torus minus a disk, a Klein bottle minus a disk, or a Möbius band minus a disk. In each case, the boundary curve(s) of $C(c \cup a)$ must bound disks in S. (Otherwise S would be a connected sum.) The case of a Klein bottle can not occur as the Klein bottle is not prime. Thus S is a torus or a projective plane.

By choosing a polygonal representation of S with the smallest number of edges we establish the lemma. □

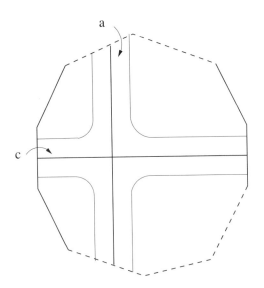

Figure 2.13. *The curve c and arc a.*

Theorem 2.2.7 (The Classification of Surfaces). *Every closed connected surface is homeomorphic to either a sphere or a connected sum of tori or a connected sum of projective planes.*

Definition 2.2.8. For an orientable surface, the number of tori in this connected sum is called the *genus*. For a non-orientable surface the number of projective planes in this connected sum is called the *genus*.

The proof of Theorem 2.2.7 duplicates the reasoning in the proof of Lemma 2.2.6. However, instead of concluding, at certain points, as we could there, that S is a torus or a projective plane, we split off summands ($\mathbb{R}P^2$ or $\mathbb{T}^2$). The fact that this strategy works, i.e., that the procedure terminates, follows from considering the number of vertices in the polygonal representation of the surface as summands are split off. In Exercise 3 below, you will prove that the number of vertices decreases as tori and projective planes are split off.

The fact that an orientable surface is built from torus summands (and not projective planes) is immediate. The fact that a non-orientable surface can be built from projective planes follows from Exercise 2 below.

Corollary 2.2.9. *For S an orientable surface of genus g, $\chi(S) = 2 - 2g$. For S a non-orientable surface of genus g, $\chi(S) = 1 - g$.*

The Euler characteristic and orientability of a closed connected surface are two examples of *topological invariants* (quantities that are well-defined for a topological manifold). Thus Corollary 2.2.9 tells us that Euler characteristic and orientability are a complete set of topological invariants for surfaces.

Our goal is to understand 3-manifolds. We hope to discover and prove a theorem analogous to Theorem 2.2.7 for 3-manifolds. Two of the most basic problems toward this endeavor are the following:

Problem 2.2.10. List all possible 3-manifolds.

Problem 2.2.11. Decide whether or not two given 3-manifolds are homeomorphic.

In Chapter 3, we will discuss prime factorization of 3-manifolds. In the context of 3-manifolds, 2-spheres will take the place of the circles used with surfaces. There are other decompositions of 3-manifolds, involving surfaces other than 2-spheres that are also of interest. We will say more about this in Section 3.4.

Exercises

Exercise 1. Show that the torus is prime.

Exercise 2. Let (S, K) and T be as in the proof of Lemma 2.2.4. Use induction on the number of vertices in T to construct a polygon P in $\mathbb{R}^2$ from a finite collection of standard 2-simplices in $\mathbb{R}^3$ corresponding to the vertices in K such that the result of identifying appropriate edges of P is the surface S.

Exercise 3. Show that in the proof of Lemma 2.2.6, $C(c \cup a)$ is either a punctured torus, punctured Klein bottle, or punctured Möbius band.

Exercise 4. Prove that $\mathbb{T}^2 \# \mathbb{R}P^2 = \mathbb{R}P^2 \# \mathbb{R}P^2 \# \mathbb{R}P^2$.

Exercise 5. Prove that if P is a polygonal representation of a surface S and $S = S_1 \# S_2$ is a non-trivial connected sum, then S_1 and S_2 admit polygonal representations whose polygons have fewer vertices than P.

2.3. Decompositions of Surfaces

In the preceding section we saw that every surface admits a decomposition into its prime summands. Note that the prime decomposition of a surface given by Theorem 2.2.7 is not unique up to isotopy, though for orientable surfaces the summands are unique up to homeomorphism. See Figures 2.14 and 2.15. Specifically, for non-orientable surfaces there are other prime decompositions than those in Theorem 2.2.7, namely those involving tori, e.g., $\mathbb{T}^2 \# \mathbb{R}P^2 = \mathbb{R}P^2 \# \mathbb{R}P^2 \# \mathbb{R}P^2$.

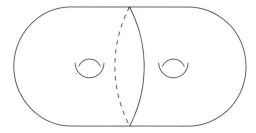

Figure 2.14. *A prime decomposition of the genus 2 surface.*

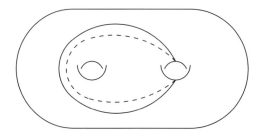

Figure 2.15. *Another prime decomposition of the genus 2 surface.*

There are other natural decompositions of surfaces. Here we will consider decompositions into pairs of pants. The thrice-punctured sphere, or *pair of pants*, forms an important building block for orientable surfaces. Of interest here is the fact that a pair of pants contains no essential simple closed curves. You will establish this property in the exercises.

Definition 2.3.1. A *pants decomposition* of a compact surface S is a collection of pairwise disjoint simple closed curves $\{c_1, \ldots, c_n\}$ such that each component of

$$S \backslash (c_1 \cup \cdots \cup c_n)$$

is a pair of pants. Two pants decompositions are *equivalent* if they are isotopic.

Pants decompositions of surfaces are not unique. See Figures 2.16 and 2.17. In the exercises you will show that every compact orientable surface of negative Euler characteristic of genus g with b boundary components admits a pants decomposition and each pants decomposition contains exactly $3g - 3 + b$ simple closed curves.

2.4. Covering Spaces and Branched Covering Spaces

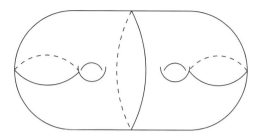

Figure 2.16. *A pants decomposition of the genus 2 surface.*

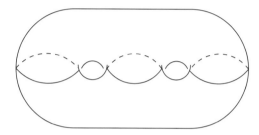

Figure 2.17. *Another pants decomposition of the genus 2 surface.*

Exercises

Exercise 1. Prove that a pair of pants contains no essential simple closed curves.

Exercise 2. Show that every compact orientable surface of negative Euler characteristic of genus g with b boundary components admits a pants decomposition and that every such pants decomposition contains exactly $3g - 3 + b$ simple closed curves.

Exercise 3*. Parametrize the isotopy classes of simple closed curves on the torus by pairs of integers.

Hint: Consider the curve $\mathbb{S}^1 \times$ (point), called the *meridian*, and the curve (point) $\times \mathbb{S}^1$, called the *longitude*.

2.4. Covering Spaces and Branched Covering Spaces

When we discussed the torus in Chapter 1, we introduced it as a quotient space of $\mathbb{R}^n$. Underlying this description of the torus is a more general notion:

Definition 2.4.1. Let $p : E \to B$ be a continuous map. The open set $U \subset B$ is said to be *evenly covered* if $p^{-1}(U)$ is a disjoint union of open sets $\{U_\alpha\}$ such that $\forall \alpha \; p|_{U_\alpha} : U_\alpha \to U$ is a homeomorphism.

Let E be a manifold, B a connected manifold, and $p : E \to B$ a continuous map. The triple (E, B, p) is a *covering* if $\forall b \in B$ there is an open set $U \subset B$ with $b \in U$ that is evenly covered.

The map p is called the *covering map*.

Example 2.4.2. The triple $(\mathbb{R}, \mathbb{S}^1, p)$, where $p : \mathbb{R} \to \mathbb{S}^1$ is given by $p(x) = e^{2\pi i x}$, is a covering. See Figure 2.18.

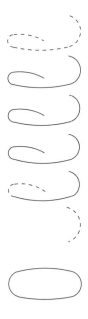

Figure 2.18. $\mathbb{R}$ *covers* $\mathbb{S}^1$.

Example 2.4.3. The triple $(\mathbb{R}^2, \mathbb{T}^2, p)$, where $p : \mathbb{R}^2 \to \mathbb{T}^2$ is given by $p(x, y) = (e^{2\pi i x}, e^{2\pi i y})$, is a covering (cf. Section 1.1).

Example 2.4.4. The triple $(\mathbb{S}^n, \mathbb{R}P^n, p)$, where $p : \mathbb{S}^n \to \mathbb{R}P^n$ is defined by $p(\mathbf{x}) = [\mathbf{x}] = [-\mathbf{x}]$ (the equivalence class containing $\mathbf{x}$ and $-\mathbf{x}$), is a covering (cf. Section 1.1).

Note that if $p : E \to B$ is a covering map, then p is a local homeomorphism, but not conversely. For example, consider $p : \mathbb{R}^+ \to \mathbb{S}^1$ given by $p(x) = e^{2\pi i x}$. Then p is a local homeomorphism but not a covering map.

Definition 2.4.5. Let (E, B, p) be a covering space. A homeomorphism $t : E \to E$ is a *covering transformation* if $p \circ t = p$.

Example 2.4.6. For $n \in \mathbb{Z}$ the map $t : \mathbb{R} \to \mathbb{R}$ given by $t(x) = x + n$ is a covering transformation for the covering in Example 2.4.2.

2.4. Covering Spaces and Branched Covering Spaces

Example 2.4.7. For $n, m \in \mathbb{Z}$ the map $t : \mathbb{R}^2 \to \mathbb{R}^2$ given by $t(x, y) = (x + n, y + m)$ is a covering transformation for the covering in Example 2.4.3.

Example 2.4.8. The antipodal map of $\mathbb{S}^n$ is a covering transformation for the covering in Example 2.4.4.

You will show in the exercises that the covering transformations for a given covering space form a group. It is called the *group of covering transformations*.

Example 2.4.9. The group of covering transformations for the covering space in Example 2.4.4 is $\mathbb{Z}/2$. Indeed, only the identity map and the antipodal map are covering maps. To see this, note that if t is a covering transformation, then $\forall \mathbf{x} \in \mathbb{S}^n$, there are only two options: either $t(\mathbf{x}) = \mathbf{x}$ or $t(\mathbf{x}) = -\mathbf{x}$. Since t is continuous, these two options extend continuously over $\mathbb{S}^n$ to either the identity map or the antipodal map.

You will show in the exercises that the group of covering transformations for the covering spaces in Examples 2.4.2 and 2.4.3 are $\mathbb{Z}$ and $\mathbb{Z}^2$, respectively.

Lemma 2.4.10. *If (E, B, p) is a covering, B is a compact connected simplicial complex, and for $b \in B$, $p^{-1}(b)$ is finite, then*
$$\chi(E) = \#\{p^{-1}(b)\} \cdot \chi(B).$$

The proof of this lemma requires showing that $\#\{p^{-1}(b)\}$ is constant and then counting simplices.

There is a more general notion:

Definition 2.4.11. Let E, B be manifolds, E' a submanifold of E, B' a submanifold of B, and $p : E \to B$ a continuous map. The quintet (E, E', B, B', p) is a *branched covering* if

- $p|_{E \setminus E'} : E \setminus E' \to B \setminus B'$ is a covering map;
- $p|_{E'} : E' \to B'$ is a covering map.

Here B' is called the *branch locus* and E' is called the *ramification locus*.

Example 2.4.12. The quintet $(\mathbb{C}, \mathbf{0}, \mathbb{C}, \mathbf{0}, f)$, where $f(\mathbf{z}) = \mathbf{z}^n$, is a branched covering.

Example 2.4.13. The quintet $(\mathbb{T}^2, \{a, b, c, d\}, \mathbb{S}^2, \{a', b', c', d'\}, p)$ can be made into a branched covering by appropriately defining p. See Figure 2.19. In Figure 2.19 an axis intersects the torus in four points, a, b, c, d. An order 2 rotation of the torus around the axis identifies the upper half of the torus with the lower half. In this identification, points are identified pairwise

except for the four points of intersection with the axis. The quotient is a sphere. We denote the quotient map by p and denote $p(a) = a'$, $p(b) = b'$, $p(c) = c'$, $p(d) = d'$. Then
$$p|_{\mathbb{T}\setminus\{a,b,c,d\}} : \mathbb{T}\setminus\{a,b,c,d\} \to \mathbb{S}^2\setminus\{a',b',c',d'\}$$
is a 2-fold cover and
$$p|_{\{a,b,c,d\}} : \{a,b,c,d\} \to \{a',b',c',d'\}$$
is a homeomorphism.

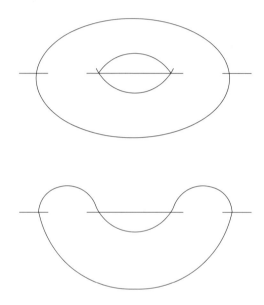

Figure 2.19. $\mathbb{T}^2$ as a branched cover of $\mathbb{S}^2$.

Lemma 2.4.14 (Riemann-Hurwitz). *If (E, E', B, B', p) is a branched covering, B is a compact connected surface, E', B' are finite sets, and for $b \in B\setminus B'$, $p^{-1}(b)$ is finite, then*
$$\chi(E) = \#\{p^{-1}(b)\} \cdot (\chi(B) - \chi(B')) + \#\{E'\}.$$

Exercises

Exercise 1. Show that the covering transformations for a given covering space form a group.

Exercise 2. Show that the group of covering transformations for the covering space in Example 2.4.2 is $\mathbb{Z}$.

Exercise 3. Show that the group of covering transformations for the covering space in Example 2.4.3 is $\mathbb{Z}^2$.

2.5. Homotopy and Isotopy on Surfaces

Exercise 4. Show that there is a covering map $p : \mathbb{T}^2 \to \mathbb{K}^2$ and that the group of covering transformations of the corresponding covering space is $\mathbb{Z}/2$.

Exercise 5. Prove the Riemann-Hurwitz formula.

2.5. Homotopy and Isotopy on Surfaces

Isotopies are always homotopies. The converse is not true in general. In this section we will see that on surfaces, homotopy classes of simple closed curves correspond to isotopy classes of simple closed curves.

Definition 2.5.1. Let F be a surface and let $\mathcal{C} = \{c_1, \ldots, c_n\}$ be a collection of simple closed curves in F that have been isotoped to intersect in a minimal number of points. We say that $\mathcal{C}$ is *filling* if $F \setminus (c_1 \cup \cdots \cup c_n)$ is a union of disks.

Example 2.5.2. The pair of curves on $\mathbb{T}^2$ pictured in Figure 2.20 is filling.

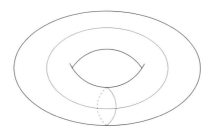

Figure 2.20. *A pair of filling curves in $\mathbb{T}^2$.*

Lemma 2.5.3 (Alexander Trick). *Suppose that $f : \mathbb{D}^n \to \mathbb{D}^n$ is a homeomorphism such that $f|_{\partial \mathbb{D}^n}$ is the identity. Then f is isotopic to the identity.*

Proof. The isotopy $H : \mathbb{D}^n \times I \to \mathbb{D}^n$ is given by

$$H(\mathbf{x}, t) = \begin{cases} tf(\mathbf{x}/t) & \text{if } 0 \leq \|\mathbf{x}\| < t, \\ \mathbf{x} & \text{if } t \leq \|\mathbf{x}\| \leq 1. \end{cases}$$

See Figure 2.21. □

Theorem 2.5.4. *Suppose that F is a closed orientable surface and $h : F \to F$ is a homeomorphism. If h is homotopic to the identity, then h is isotopic to the identity.*

The key ingredient in this proof is that homotopic simple closed curves in F are isotopic. We provide a sketch of the argument in a special case.

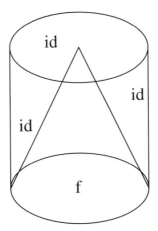

Figure 2.21. *Schematic for the Alexander Trick.*

Proof in the case $F = \mathbb{T}^2$. Let $h : F \to F$ be a homeomorphism that is homotopic to the identity. Let c be a simple closed curve in $\mathbb{T}^2$ and compare c and $h(c)$. We may assume that $c \pitchfork h(c)$. See Figure 2.22.

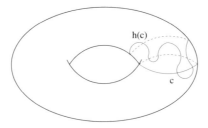

Figure 2.22. *Curves c and $h(c)$.*

If there is a disk that is cobounded by an arc in c and an arc in $h(c)$, then we can use an innermost such disk to perform an isotopy that reduces the number of points of intersection between c and $h(c)$. We continue to perform isotopies reducing the number of points of intersection until there are no disks cobounded by an arc in c and an arc in $h(c)$. See Figure 2.23.

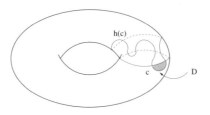

Figure 2.23. *An innermost disk D.*

2.6. The Mapping Class Group

Now every component of $h(c)\setminus c$ is either entirely contained in $\mathbb{T}^2\setminus c$ or crosses the open annulus $\mathbb{T}^2\setminus c$. In particular, the orientations of the points in $c \cap h(c)$ are the same. Recall that $[c] \cdot [c] = 0$. By Theorem 2.1.7, $[c] \cdot [h(c)] = 0$. It follows that $c \cap h(c) = \emptyset$. Thus $h(c)$ is contained in the annulus $\mathbb{T}^2\setminus c$. Therefore, $h(c)$ is isotopic to c, whence $h|_c$ is isotopic to $\mathrm{id}|_c$.

Consider the pair of filling curves $\{m, l\}$ in Example 2.5.2. The above argument allows us to assume that, after isotopy, $h(m)$ coincides with m. You will prove in the exercises that there is an isotopy fixing m after which l and $h(l)$ coincide, i.e., that $h|_l$ is isotopic to $\mathrm{id}|_l$.

Now consider the polygonal representation of $\mathbb{T}^2$ with four sides that identify to $m \cup l$. Then h defines a map on this square Q and $h|_{\partial Q} = \mathrm{id}_{\partial Q}$. By the Alexander Trick, the map $h|_Q$ is isotopic to the identity. Thus h is isotopic to the identity. (This completes the proof in the case of the torus.) $\square$

Exercises

Exercise 1. For m, l, h as in the proof of Theorem 2.5.4, prove that there is an isotopy fixing m after which l and $h(l)$ coincide.

Exercise 2. Find a filling pair of curves for the genus 2 surface.

Exercise 3. Show that on any surface, homotopic simple closed curves are isotopic.

2.6. The Mapping Class Group

Self-diffeomorphisms of surfaces are interesting in their own right but are of particular interest here because of their role in certain constructions on 3-manifolds (e.g., mapping tori (see Chapter 3) and the curve complex (see Chapter 7)). More specifically, it is not self-diffeomorphisms of a surface but isotopy classes of self-diffeomorphisms of a surface that play a role. For this reason we are interested in the mapping class group defined below.

In the exercises, you will show that the set of all self-diffeomorphisms of a topological space forms a group. In the case of an oriented surface, we denote this group by $\mathrm{Diffeo}(S)$. You will also show that the set of orientation-preserving self-diffeomorphisms and the set of self-diffeomorphisms that are homotopic to the identity form normal subgroups. In the case of an oriented surface, we denote these groups by $\mathrm{Diffeo}_+(S)$ and $\mathrm{Diffeo}_0(S)$, respectively.

We will define the mapping class group in terms of diffeomorphisms. Homeomorphisms of surfaces are isotopic to diffeomorphisms; hence our definitions could equally well be made in terms of homeomorphisms. In particular, our definition of Dehn twist given below yields homeomorphisms rather than diffeomorphisms. However, these homeomorphisms can easily be smoothed.

Definition 2.6.1. Let S be a compact orientable surface. The *mapping class group* of S, denoted by $\mathcal{MCG}(S)$ is $\text{Diffeo}_+(S)/\text{Diffeo}_0(S)$, the group of orientation-preserving self-diffeomorphisms of S modulo the subgroup consisting of self-diffeomorphisms of S homotopic to the identity.

Lemma 2.6.2. *The mapping class group of the torus is isomorphic to $SL(2,\mathbb{Z})$, the group of 2×2 matrices with integer coefficients and determinant 1.*

Proof. Given a simple closed curve γ on $\mathbb{T}^2$, we define an integer vector

$$i(\gamma) = ([\gamma] \cdot [m], [\gamma] \cdot [b]).$$

As we saw in the proof of Theorem 2.5.4, the vector $i(\gamma)$ completely determines the isotopy class of γ on $\mathbb{T}^2$. Then, given a homeomorphism $f \in \text{Homeo}(\mathbb{T}^2)$, we define the matrix A_f, whose columns are the vectors $i(f(l))$ and $i(f(m))$.

The proof of Theorem 2.5.4 shows that the matrix A_f completely determines the isotopy class of f. In the exercises you will verify that the map

$$\eta : f \mapsto A_f$$

is a homomorphism from $\text{Homeo}(\mathbb{T}^2)$ to the group of integer matrices. In particular, the image of η is contained in $GL(2,\mathbb{Z})$.

Since for homotopic curves a, a' we have

$$i(a) = i(a'),$$

it follows that for $g \in \text{Homeo}_0(\mathbb{T}^2)$,

$$A_g = I,$$

the identity matrix in $GL(2,\mathbb{Z})$. Therefore, the homomorphism η descends to a homomorphism

$$H : \mathcal{MCG}(\mathbb{T}^2) \to GL(2,\mathbb{Z}).$$

Below, we will verify that H is an isomorphism onto $SL(2,\mathbb{Z})$. As a first step, given a matrix $A \in SL(2,\mathbb{Z})$,

$$A = \begin{bmatrix} a & c \\ b & d \end{bmatrix},$$

we construct a homeomorphism $f = f_A \in \text{Homeo}(\mathbb{T}^2)$ such that $A_f = A$ as follows: Recall the covering map $p : \mathbb{R}^2 \to \mathbb{T}^2$ from Example 2.4.3 and consider the matrix A as a linear transformation of $\mathbb{R}^2$. Then A maps horizontal lines $p^{-1}(m)$ to lines of slope $\frac{b}{a}$ and these project to simple closed curves α on $\mathbb{T}^2$ so that

$$i(\alpha) = \begin{bmatrix} a \\ b \end{bmatrix}.$$

Likewise, A maps vertical lines $p^{-1}(l)$ to lines of slope $\frac{d}{c}$ and these project to simple closed curves β on S so that

$$i(\beta) = \begin{bmatrix} c \\ d \end{bmatrix}.$$

We define a map $f = f_A : \mathbb{T}^2 \to \mathbb{T}^2$ as follows: Let $x \in \mathbb{T}^2$ and let $\tilde{x}$ be a point in $\mathbb{R}^2$ such that $p(\tilde{x}) = x$. Set $f(x) = p \circ A(\tilde{x})$. You will verify in the exercises that the map $f : \mathbb{T}^2 \to \mathbb{T}^2$ is well-defined and is a homeomorphism. By construction,

$$i(f(l)) = \begin{bmatrix} a \\ b \end{bmatrix}, \quad i(f(m)) = \begin{bmatrix} c \\ d \end{bmatrix}.$$

Therefore, $A_f = A$. Note that, in the charts of $\mathbb{T}^2$ implicitly defined by the covering map p, the Jacobian matrix of f is the matrix A. Thus, if $A \in SL(2, \mathbb{Z})$, then $f = f_A$ is orientation-preserving. Therefore, it immediately follows that

$$SL(2, \mathbb{Z}) \subset \text{im}(H).$$

Furthermore, if $A = A_g, g \in \text{Homeo}(\mathbb{T}^2)$, then for $f = f_A$, we obtain

$$A = A_f = A_g$$

and, hence, the maps f_A and g are isotopic to each other. Therefore, the map

$$J : A \mapsto [f_A] \in \mathcal{MCG}(\mathbb{T}^2)$$

satisfies

$$J \circ H = \text{id}.$$

Hence, the homomorphism H is injective. It remains to show that $\text{im}(H) \subset SL(2, \mathbb{Z})$. Note that each diffeomorphism $f = f_A$ for which $\det(A) = -1$ is orientation-reversing. Thus, if $g \in \text{Homeo}(\mathbb{T}^2)$ is such that $\det(A_g) = -1$, then g is isotopic to the orientation-reversing diffeomorphism $f = f_A$. Since an orientation-preserving diffeomorphism cannot be isotopic to an orientation-reversing one, it follows that for each $g \in \text{Homeo}_+(\mathbb{T}^2)$,

$$A_g \in SL(2, \mathbb{Z}).$$

This concludes the proof that H is an isomorphism $\mathcal{MCG}(\mathbb{T}^2) \to SL(2, \mathbb{Z})$.

□

Remark 2.6.3. The reader familiar with homology will recognize that the map i defines coordinates on the homology group $H_1(\mathbb{T}^2, \mathbb{Z})$, which is isomorphic to $\mathbb{Z}^2$. Furthermore, A_f is the marix of the automorphism of $H_1(\mathbb{T}^2, \mathbb{Z})$ determined by f.

A simple closed curve c in a surface S has a *regular neighborhood* (for the technical definition of regular neighborhood, see Appendix A). When c is orientation-preserving, this means that there is an embedding of an annulus $f : \mathbb{S}^1 \times [0,1] \to S$ such that $c = f(\mathbb{S}^1 \times \{1/2\})$ whose image is the regular neighborhood. The regular neighborhood can be taken to be open or closed. If it is open, we write $\eta(c)$ (for $f(\mathbb{S}^1 \times (0,1))$); if it is closed, we write $N(c)$ (for $f(\mathbb{S}^1 \times [0,1])$). If c is orientation-reversing, an analogous statement holds with the annulus replaced by a Möbius band.

Definition 2.6.4. Let c be an orientation-preserving simple closed curve in a compact surface S and let $N(c)$ be a regular neighborhood of c that is oriented via the parametrization $i : \mathbb{S}^1 \times [0,1] \to \eta(c)$. A map $f : S \to S$ is called a *left Dehn twist* around c if

- $f|_{S \setminus N(c)}$ is the identity map; and
- $f|_{N(c)}$ is the map of the annulus given by

$$f(e^{2i\pi\theta}, t) = (e^{2i\pi(\theta+t)}, t) \quad \forall (e^{2i\pi\theta}, t) \in \mathbb{S}^1 \times [0,1].$$

See Figure 2.24. Likewise, a map $f : S \to S$ is called a *right Dehn twist* around c if

- $f|_{S \setminus N(c)}$ is the identity map; and
- $f|_{N(c)}$ is the map of the annulus given by

$$f(e^{2i\pi\theta}, t) = (e^{2i\pi(\theta-t)}, t) \quad \forall (e^{2i\pi\theta}, t) \in \mathbb{S}^1 \times [0,1].$$

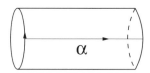 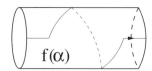

Figure 2.24. *Local effect of a left Dehn twist.*

We can think of a (left/right) Dehn twist as doing the following: cutting the surface along the core of the annulus, performing a full twist (to the left/to the right), and regluing. Theorem 2.6.5 shows that Dehn twists are the basic building blocks for all surface diffeomorphisms.

Theorem 2.6.5 (Dehn, 1938; Lickorish, 1962). *Every surface diffeomorphism is isotopic to a composition of Dehn twists. In other words, the mapping class group is generated by Dehn twists.*

2.6. The Mapping Class Group

Theorem 2.6.5 was proved by M. Dehn and discussed in his Breslau Lectures in 1922. It was published in 1938; see [**33**]. The theorem was rediscovered (via a rather clever argument that we outline below) by W. B. R. Lickorish in 1962; see [**88**]. We will discuss some of the insights that go into the proof of this theorem.

Lemma 2.6.6. *If the simple closed curves α, β in the compact surface S intersect exactly once, then there is a pair f_1, f_2 of Dehn twists such that $f_2 \cdot f_1(\alpha)$ is isotopic to β.*

Proof. If we choose f_1 to be a left Dehn twist along β and f_2 to be a right Dehn twist along α, then indeed $f_2 \cdot f_1(\alpha) = \beta$. See Figures 2.25, 2.26, and 2.27. □

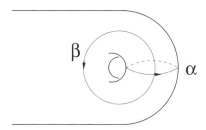

Figure 2.25. α and β.

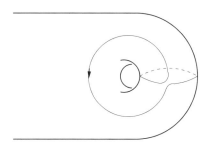

Figure 2.26. $f_1(\alpha)$.

Lemma 2.6.7. *If α, β are oriented simple closed curves in the orientable surface S, then there is a series of Dehn twists $f_1, f_2, \ldots, f_k$ such that $f_k \cdot \ldots \cdot f_2 \cdot f_1(\alpha)$ is either disjoint from β or intersects β in exactly two oppositely oriented points.*

Sketch of proof. The proof is by induction on the number of points in the intersection of the two curves. Suppose that there are two points of intersection with the same orientation that are adjacent along β. See Figure 2.28. Then denote the subarc of α that connects these two points by $\tilde{c}$. By

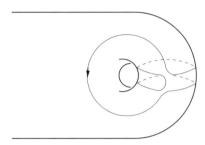

Figure 2.27. $f_2 \cdot f_1(\alpha)$.

taking the union of the subarc (or one such subarc if there are two) of β along which these two points are adjacent (either will do if there are two) with $\tilde{c}$, we construct a simple closed curve. After a small isotopy, this curve intersects α in exactly one point. Denote this curve by c. In Exercise 3 you will show that for f_1 a right Dehn twist along c, $f_1(\alpha)$ has fewer points of intersection with β than α.

If there are no points of intersection with the same orientation that are adjacent along β and if there are at least three points of intersection, then a similar construction allows a reduction of the number of points in $\alpha \cap \beta$. You will show this in Exercise 4. □

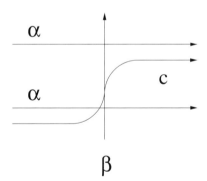

Figure 2.28. *Adjacent points of intersection of α and β.*

To prove Theorem 2.6.5, let S be a closed orientable surface. There is a standard collection of curves $a_1, b_1, \ldots, a_g, b_g$ that cuts S into a disk; see Figure 2.29. A self-diffeomorphism of S takes this fixed collection of curves to another such collection of curves. Using Lemmas 2.6.6 and 2.6.7 allows us to perform a series of Dehn twists that reverses the effect of the surface diffeomorphism on the specified collection of curves. The collection of Dehn twists can be seen as "undoing" the effect of the surface diffeomorphism on

2.6. The Mapping Class Group

the specified collection of curves. The map on the complementary disk is isotopic to the identity by the Alexander Trick.

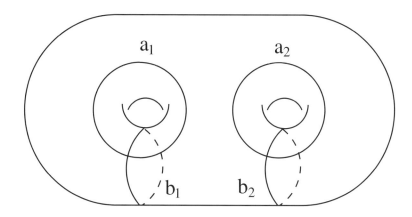

Figure 2.29. *Standard collection of curves a_1, b_1, a_2, b_2 in a genus 2 surface.*

Definition 2.6.8. A diffeomorphism $h : S \to S$ is called *periodic* if $h^n = \mathrm{id}_S$ for some $n \in \mathbb{N}$. It is called *reducible* if $\exists$ an essential simple closed curve $C \subset S$ such that $\phi(C) = C$ (setwise).

Example 2.6.9. Let $h_1 : \mathbb{T}^2 \to \mathbb{T}^2$ be defined by $h_1(e^{i\theta}, e^{i\phi}) = (e^{-i\phi}, e^{i\theta})$. Then h_1 is periodic; in fact, $h_1^4 = \mathrm{id}$.

Let $h_2 : \mathbb{T}^2 \to \mathbb{T}^2$ be defined by $h_2(e^{i\theta}, e^{i\phi}) = (e^{i\theta}, e^{i(\theta+\phi)})$. Then h_2 is reducible, and it preserves the essential simple closed curve $\{(1, e^{i\phi}) \mid 0 \leq \phi < 2\pi\}$. More generally, for any surface S a Dehn twist along an essential simple closed curve is reducible.

Let $h_3 : \mathbb{T}^2 \to \mathbb{T}^2$ be defined by $h_3(e^{i\theta}, e^{i\phi}) = (e^{i(2\theta+\phi)}, e^{i(\theta+\phi)})$. Then h_3 is neither periodic nor reducible.

More generally,

Definition 2.6.10. An element of the mapping class group of S is *periodic* if it has a periodic representative. An element of the mapping class group of S is *reducible* if it has a reducible representative.

It is a deep theorem due to W. Thurston that elements of the mapping class group are classified as being either periodic, reducible, or pseudo-Anosov. We discuss what it means to be pseudo-Anosov in Chapter 7, when we discuss laminations.

Exercises

Exercise 1. Let X be a topological space. Show that the set of self-diffeomorphisms of X forms a group.

Exercise 2. Let S be an oriented surface. Show that $\text{Diffeo}_+(S) \lhd \text{Diffeo}(S)$ and $\text{Diffeo}_0(S) \lhd \text{Diffeo}(S)$.

Exercise 3. Show that the map $f_A : \mathbb{T}^2 \to \mathbb{T}^2$ constructed in the proof of Lemma 2.6.2 is well-defined and is a self-homeomorphism of the torus. (Hint: First, show that for every covering transformation $\tau \in \mathbb{Z}^2$ of the covering p, there exists another covering transformation τ' so that
$$A \circ \tau = \tau' \circ A.$$
This will imply that the map f_A is well-defined. Then, for $B = A^{-1} \in SL(2, \mathbb{Z})$ verify that $f_A = f_B^{-1}$.)

Exercise 4. Let α, β be as in Lemma 2.6.7, except that now there is one point of $\alpha \cap \beta$ with the opposite orientation between the two points with the same orientation. Prove that a similar construction still reduces the number of points in $\alpha \cap \beta$. See Figure 2.30.

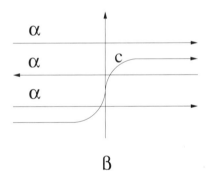

Figure 2.30. *Adjacent points of intersection of α and β.*

Exercise 5. Complete the proof of Theorem 2.6.5 in the case of the torus. (Hint: Let m, l be the curves considered in the proof of Lemma 2.6.2. Let $h : \mathbb{T}^2 \to \mathbb{T}^2$ be a homeomorphism. Construct a homeomorphism $\hat{h}$, realized by a product of Dehn twists, such that $\hat{h}^{-1} \circ h(m) = m$. Conclude that $\hat{h}^{-1} \circ h$ corresponds to the matrix
$$A = \begin{bmatrix} 1 & * \\ 0 & 1 \end{bmatrix}$$
and is hence a Dehn twist around m.)

Chapter 3

3-Manifolds

One of our goals in this chapter is to discuss prime factorization of 3-manifolds. This prime factorization is almost unique, but the 2-sphere submanifolds of a 3-manifold M along which M decomposes can be positioned in more than one way. We will also discuss other decompositions of 3-manifolds. For more on these subjects, see [**63**], [**67**], [**71**], [**73**], and [**72**].

In this chapter and beyond, we will assume familiarity with the concept of *general position*. For details, see Appendix A. One of the insights gleaned from general position is that compact 2-dimensional submanifolds of a 3-manifold M can always be isotoped slightly to intersect in a compact 1-dimensional submanifold. See Figure 3.1. A compact 1-dimensional submanifold and a compact 2-dimensional submanifold of M can always be isotoped slightly to intersect in a finite number of points. See Figure 3.2. Compact 1-dimensional submanifolds of M can be isotoped to be disjoint. A point, i.e., a 0-dimensional submanifold of M, can be assumed to be disjoint from 2-, 1-, or other 0-dimensional submanifolds of M.

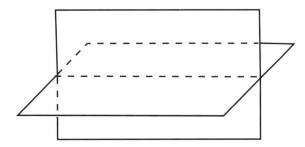

Figure 3.1. *Local picture of general position for two 2-submanifolds of a 3-manifold.*

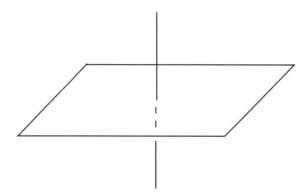

Figure 3.2. *Local picture of general position for a 1-submanifold and a 2-submanifold of a 3-manifold.*

More generally, let M be an n-manifold containing a k-dimensional submanifold K and an l-dimensional submanifold L that are in general position. Then $K \cap L$ is a (possibly empty) submanifold of M of dimension $(k+l-n)$. Furthermore, general position can always be guaranteed and maintained by small isotopies.

Convention 3.0.1. To simplify our statements, we will, in this section and beyond, make the assumption that, for $k \leq n$, a manifold of dimension k in a manifold of dimension n is a submanifold. In particular, simple closed curves and simple arcs lying in a surface or 3-manifold and surfaces lying in a 3-manifold are submanifolds.

An interesting 2-sphere contained in $\mathbb{R}^3$ that is not a submanifold in the sense defined in Chapter 1 was exhibited by J. W. Alexander. It is called the *Alexander horned sphere*. (For an artistic rendition, see for instance [**127**].) An interesting fact concerning the Alexander horned sphere is that it does not bound a 3-ball. Convention 3.0.1 now eliminates this type of example from our discussion.

3.1. Bundles

Recall two of the examples of surfaces with boundary discussed in Chapter 2: the annulus and the Möbius band. See Figures 3.3, 3.4, 3.5, and 3.6.

Figure 3.3. *The annulus.*

3.1. Bundles

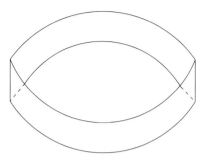

Figure 3.4. *The annulus.*

Figure 3.5. *The Möbius band.*

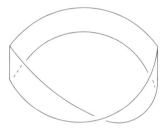

Figure 3.6. *The Möbius band.*

Definition 3.1.1. A *bundle* is a quartet (M, F, B, π) where M, F, and B are manifolds and $\pi : M \to B$ is a continuous map such that the following hold:

- for every $b \in B$, $\pi^{-1}(b)$ is homeomorphic to F;
- there is an atlas $\{U_\alpha\}$ for B such that $\forall \alpha \; \exists$ a homeomorphism $h_\alpha : \pi^{-1}(U_\alpha) \to U_\alpha \times F$;
- the following diagram commutes (the right arrow is projection onto the first factor):

$$\begin{array}{ccc} \pi^{-1}(U_\alpha) & \xrightarrow{h_\alpha} & U_\alpha \times F \\ \pi \downarrow & & \downarrow \\ U_\alpha & \xrightarrow{\text{id}} & U_\alpha. \end{array}$$

Here M is called the *total space*, F is called the *fiber*, B is called the *base space*, and π is called the *projection*. A pair (U_α, h_α) is called a *bundle chart*. The family $\{(U_\alpha, h_\alpha)\}$ is called a *bundle atlas*.

We sometimes refer to (M, F, B, π) simply as an F-bundle over B.

In the context of 3-manifolds, we sometimes use the word *fibering* instead of bundle.

Example 3.1.2. Any product manifold $X \times Y$ is a bundle. The total space is $X \times Y$, the fiber is Y (or X, respectively), the base space is X (or Y, respectively), and the projection is projection onto the first (or second, respectively) factor.

The annulus is an example of this type of bundle.

Example 3.1.3. The Möbius band is a bundle with fiber I and base space $\mathbb{S}^1$. The projection can be described by considering the rectangle in Figure 3.5. The projection consists in vertical projection onto the core curve, that is, the marked curve. We leave it as an exercise to show that the map thus obtained is continuous, as required.

Example 3.1.4. Let S be a closed connected n-manifold and $f : S \to S$ a homeomorphism. The *mapping torus* of f is the $(n+1)$-manifold obtained from $S \times [-1, 1]$ by identifying the points $(x, -1)$ and $(f(x), 1)$, $\forall x \in S$. The mapping torus of f is a bundle with fiber S and base space $\mathbb{S}^1$. The total space is often denoted by $(S \times I)/\sim_f$.

The role of the homeomorphism f in the above example is an interesting one. One crucial fact is that the isotopy class of this homeomorphism determines the $(n+1)$-manifold.

The Möbius band is an example of a mapping torus, with $S = [-1, 1]$ and $f : [-1, 1] \to [-1, 1]$ given by $f(x) = -x$.

Definition 3.1.5. Suppose that (M, F, B, π) and (M', F, B', π') are bundles. An *isomorphism* between the bundles is a pair of homeomorphisms

$$h : M \to M'$$

and

$$f : B \to B'$$

such that the following diagram commutes:

$$\begin{array}{ccc} M & \xrightarrow{h} & M' \\ \pi \downarrow & & \downarrow \pi' \\ B & \xrightarrow{f} & B'. \end{array}$$

We say that two bundles are *isomorphic* or *equivalent* if there is an isomorphism between them.

The annulus and the Möbius band are inequivalent bundles. To see this, note, for instance, that the total space of the annulus has two boundary components whereas the total space of the Möbius band has only one. Thus there can be no bundle homeomorphism between the two.

There is a more intuitive approach to understanding the difference between the annulus and the Möbius band: Cut the total spaces along their core curves. The core curve of the annulus is marked in Figure 3.3; that of the Möbius band is marked in Figure 3.5. If we cut along the core curve of the annulus, we obtain two annuli. If we cut along the core curve of the Möbius band, we obtain only one annulus. Both the annulus and the Möbius band are *I*-bundles over the circle, in fact, they are the only two *I*-bundles over the circle.

Example 3.1.6. The annulus and the Möbius band are inequivalent bundles.

Definition 3.1.7. A bundle that is isomorphic to a product bundle is called a *trivial bundle*. A bundle that is not a trivial bundle is called a *non-trivial bundle*.

Example 3.1.8. The annulus is a trivial bundle. The Möbius band is a non-trivial bundle.

Definition 3.1.9. Given a bundle $E = (M, F, B, \pi)$ and a submanifold B' of B, the *restriction* of E to B' is a bundle with total space $M' = \pi^{-1}(B')$, fiber F, base space B', and projection $\pi|_{M'}$. We denote this restriction by $E|_{B'}$.

Example 3.1.10. We construct a non-trivial *I*-bundle over the Möbius band as follows: Consider the rectangle in Figure 3.5. Points in the rectangle lie in $[-1, 1] \times [-1, 1]$. To form the Möbius band, we identify $(-1, x)$ with $(1, -x)$. Now consider $[-1, 1] \times [-1, 1] \times [-1, 1]$ and identify $(-1, x, y)$ with $(1, -x, -y)$. The result is a non-trivial bundle over the Möbius band. We leave it as an exercise to show that the total space of this bundle is a solid torus. See Figure 3.7.

The above example can be visualized as follows: First we visualize the Möbius band. We start with a square and identify the left and right sides so that we obtain a non-trivial bundle over the core circle. We do so by stretching and bending the square so that the left and right sides of the square come up out of the page. We then twist the right side by 180 degrees and attach the two sides.

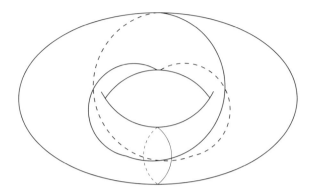

Figure 3.7. *The solid torus containing the Möbius band.*

Now we visualize the non-trivial bundle over the Möbius band. We start with the trivial *I*-bundle over the square as in Figure 3.5. This is a cube. We identify the left and right sides of the cube so that we obtain a non-trivial bundle. We do so by stretching and bending the cube so that the two squares come up out of the page. We must identify these squares so that (if we put our head between the two and look to one side and then the other) the left side of one is identified to the left side of the other (and the right side of one to the right side of the other) and the top of one is identified to the bottom of the other.

The trivial bundle $\mathbb{B}^2 \times \mathbb{S}^1$ and the bundle in Example 3.1.10 are inequivalent bundles. Their total spaces are homeomorphic, but their bundle structures differ.

Definition 3.1.11. A *section* of a bundle is a continuous map $\sigma : B \to E$ such that $\pi \circ \sigma = \mathrm{id}_B$ (where id_B is the identity map on B).

For a comprehensive treatment of bundles, see [**65**] or [**150**].

Definition 3.1.12. Let M be an n-manifold. Let S be a submanifold of M of dimension m. A *regular neighborhood* of S is a submanifold $N(S)$ of M of dimension n that is the total space of a bundle over S with fiber $\mathbb{B}^{n-m}$. A regular neighborhood of a 1-manifold in a 3-manifold is also called a *tubular* neighborhood. A regular neighborhood of an $(n-1)$-dimensional submanifold of an n-manifold that is a trivial bundle is also called a *collar* (or, if $n = 2$, a *bicollar*). An *open regular neighborhood* of S is a submanifold $\eta(S)$ of M of dimension n that is the total space of a bundle over S with fiber $interior(\mathbb{B}^{n-m})$.

Figures 3.8 and 3.9 exhibit the local picture of regular neighborhoods of simple closed curves, simple arcs, and surfaces in a 3-manifold. For more

3.1. Bundles

Figure 3.8. *Local picture of a 1-manifold and its regular neighborhood in a 3-manifold.*

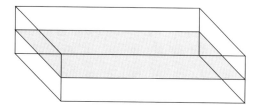

Figure 3.9. *Local picture of a surface and its regular neighborhood in a 3-manifold.*

details see Appendix A. In the exercises you will show that for simple closed curves there are only two possible total spaces for regular neighborhoods. In particular, if c is an orientation-preserving simple closed curve in an orientable 3-manifold M, then there is a solid torus, $f : \mathbb{S}^1 \times \mathbb{D}^2 \to M$, such that $c = f(\mathbb{S}^1 \times \{(0,0)\})$, whose image is the regular neighborhood of c in M. When c is an orientation-reversing simple closed curve in a non-orientable 3-manifold M, then there is an embedding of a twisted $\mathbb{D}^2$-bundle over $\mathbb{S}^1$, $f : \mathbb{S}^1 \tilde{\times} \mathbb{D}^2 \to M$, such that $c = f(\mathbb{S}^1 \tilde{\times} \{(0,0)\})$, whose image is the regular neighborhood of c in M.

The following theorem summarizes results discussed in Appendix A.

Theorem 3.1.13. *Let M be an n-manifold and let S be a k-manifold in M. There exists a regular neighborhood $N(S)$ for S in M. Furthermore, any two regular neighborhoods of S in M are isotopic.*

A related, but more specific, theorem is the following:

Theorem 3.1.14. *If S is a surface in the 3-manifold M and both M and S are orientable, then $N(S)$ is a trivial I-bundle.*

Definition 3.1.15. Let S be a surface in the 3-manifold M. We say that S is a *2-sided* surface if a regular neighborhood of S in M is a trivial I-bundle. We say that S is a *1-sided* surface if a regular neighborhood of S in M is a non-trivial, or *twisted*, I-bundle.

Exercises

Exercise 1. Prove that the projection mentioned in Example 3.1.3 is continuous (as required).

Exercise 2. Prove that the total space of the bundle described in Example 3.1.10 is the solid torus.

Exercise 3. Use the core curves of the annulus and of the Möbius band to describe sections of the bundles described in Examples 3.1.2 and 3.1.3.

Exercise 4. Prove that there are only two bundles with base space $\mathbb{S}^1$ and fiber I.

Exercise 5. Prove that there are only two bundles with base space $\mathbb{S}^1$ and fiber $\mathbb{B}^n$.

3.2. The Schönflies Theorem

The Schönflies Theorem generalizes the Jordan Curve Theorem. One of the key notions in this context is Definition 3.2.1:

Definition 3.2.1. A 3-manifold M is *irreducible* if every 2-sphere in M bounds a 3-ball. A 3-manifold is *reducible* if it contains a 2-sphere that does not bound a 3-ball.

Example 3.2.2. The 3-manifold $\mathbb{S}^2 \times \mathbb{S}^1$ is reducible.

Before stating and proving the Schönflies Theorem, we prove a lemma. We will use this lemma in the proof of the Schönflies Theorem. It will also serve as a warm-up for our discussion of 3-manifolds.

Lemma 3.2.3. *Let S_1, S_2 be surfaces in a 3-manifold M. Suppose that $S_1 \cap S_2$ contains a simple closed curve c such that, for $i = 1, 2$, one component of $S_i \backslash c$ is a disk D_i such that D_2 is disjoint from S_1. If $c \cup D_1 \cup D_2$ bounds a 3-ball in M, then*

- *$(S_1 \backslash D_1) \cup D_2$ is isotopic to S_1;*
- *there is an isotopy of S_1 that eliminates the curve c from $S_1 \cap S_2$ and introduces no new components of intersection.*

Note that we are considering S_1 before and after the isotopy. It would make sense to write S_1^t to distinguish the positioning of S_1 at various times during the isotopy, but this is usually unnecessary, as context makes it clear which positioning is meant. So we simply write S_1.

3.2. The Schönflies Theorem

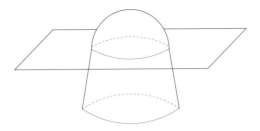

Figure 3.10. *Surfaces intersecting.*

Proof. A neighborhood of the 3-ball in M bounded by $D_1 \cup D_2$ is homeomorphic to a neighborhood of the upper hemisphere of the standard 3-ball in $\mathbb{R}^3$. See Figure 3.10, where D_1 is the disk above the xy-plane.

To obtain the first conclusion, note that straight line projection of $D_1 \subset S_1$ onto the xy-plane is an isotopy that replaces S_1 with $(S_1 \backslash D_1) \cup D_2$.

To obtain the second conclusion, note that straight line projection of the portion of S_1 lying above $y = -\epsilon$ (this includes D_1) onto the plane $y = -\epsilon$ is an isotopy of S_1 that eliminates the component c from $S_1 \cap S_2$. See Figure 3.11. □

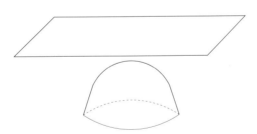

Figure 3.11. *After the projection and isotopy.*

Definition 3.2.4. Let X be an n-manifold and x a subset of X (typically a submanifold of dimension $n-1$). To *cut X along x* means to consider $X \backslash \eta(x)$. Likewise, *x cuts X into* a given set of manifolds means that $X \backslash \eta(x)$ consists of this set of manifolds.

Theorem 3.2.5 (Schönflies Theorem)**.** *Any 2-sphere in $\mathbb{R}^3$ bounds a 3-ball.*

A beautiful elementary proof of this theorem was given by Morton Brown in 1960; see [**18**]. Students are encouraged to read this proof in the original. The following proof is based on one that appeared in a lecture series given by Andrew Casson in China in 2002. We include it here because it illustrates some of the techniques employed in the contemporary study of 3-manifolds. It relies on the the Poincaré-Hopf Index Theorem mentioned in Chapter 1.

If you don't understand the proof below in its entirety on a first reading, focus on the intuition behind it and the point of view espoused. You will see the concepts and techniques used to prove this theorem over and over again. We are working in the DIFF category.

We will use the concept of a "height function". A *height function* is a Morse function with at most two critical points, namely a maximum and a minimum. This is a strong requirement. In fact, there are only three connected 3-manifolds that admit height functions: $\mathbb{S}^3$ (requires a maximum and a minimum, by compactness), $\mathbb{B}^3$ (requires a maximum and a minimum, by compactness), and $\mathbb{R}^3$ (requires no critical points). For more information on Morse functions, see Appendix B.

We will also use induction on an ordered pair of numbers. Recall that induction can be performed on any countable ordered set. When we perform induction on an ordered pair of numbers, both entries must come from a countable ordered set. The product of the countable sets is countable and is ordered via the dictionary order.

Proof. Let $S \subset \mathbb{R}^3$ be a 2-sphere submanifold. We isotope S so that the height function $h : \mathbb{R}^3 \to \mathbb{R}$ given by projection onto the third coordinate restricts to a Morse function $h|_S$.

By the Poincaré-Hopf Index Theorem,

$$\#\text{maxima} + \#\text{minima} - \#\text{saddles} = \chi(S) = 2.$$

Note that since S is compact, $h|_S$ has at least one maximum and one minimum. Let n be the number of saddle points. If $n = 0$, then S has one maximum, y_2, and one minimum, y_1. We may assume, by multiplying h with an appropriate constant, that $y_2 - y_1 = 2$. We may further assume, by translating h upwards or downwards, that $y_2 = 1$ and hence $y_1 = -1$. Each r such that $-1 < r < 1$ corresponds to a level curve in S that is a simple closed curve in the plane $z = r$. It follows from the Jordan Curve Theorem that each such curve can be isotoped within the plane $z = r$ into the circle such that $x^2 + y^2 = 1 - r^2$. It is not too hard to see (though harder to prove rigorously) that these isotopies can be chosen so as to vary smoothly with r. Thus S is isotopic to the standard 2-sphere in $\mathbb{R}^3$ and hence bounds a 3-ball.

Note that if $n = 1$, then there are two minima (or one minimum, resp.), one maximum (or two maxima, resp.), and one saddle. There are two possibilities: a "non-nested" saddle or a "nested" saddle. See Figures 3.12 and 3.13. In either case, there is a plane $z = c$ (containing the saddle) that meets

3.2. The Schönflies Theorem

S in a figure 8. Call the plane H. Then $H \cap S$ is a figure 8 that cuts S into three disks, D_1, D_2, D_3. Each of these disks contains exactly one critical point (either a maximum or a minimum). Furthermore, the figure 8 cuts H into two disks and one unbounded component. At least one of these two disks, call it D, shares its boundary with one of D_1, D_2, D_3. Attach ∂D to the boundary of the appropriate D_1, D_2, D_3, say D_1, to obtain a piecewise smooth 2-sphere $D \cup D_1$.

After a small isotopy (that introduces only one new critical point), $D \cup D_1$ has exactly two critical points (a maximum and a minimum). Since it has no saddles, the case above applies and it bounds a 3-ball. By Lemma 3.2.3, $(S \backslash D_1) \cup D$ is isotopic to S. After a small isotopy, $(S \backslash D_1) \cup D$ is smooth and has no saddles. Thus the situation reduces to the case $n = 0$.

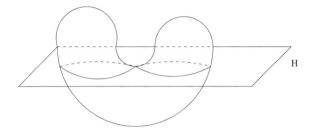

Figure 3.12. *A sphere with one (non-nested) saddle.*

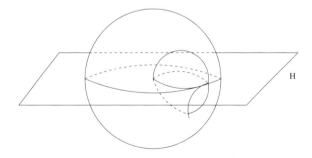

Figure 3.13. *A sphere with one (nested) saddle.*

More generally, suppose that $n \geq 2$ and that every 2-sphere in $\mathbb{R}^3$ with less than n saddles bounds a 3-ball. Then there is a plane H', given by $z = r$, for a regular value r, such that there are saddle points both above and below H' and such that the number of components of $H' \cap S$ is minimal. Denote the number of components of $H' \cap S$ by $\#|H' \cap S|$. We argue by induction on $(n, \#|H' \cap S|)$.

Here $H' \cap S$ is a compact 1-manifold (without boundary), so $H' \cap S$ is a finite union of disjoint simple closed curves in H'. Let a be an innermost component of $H' \cap S$ in H'. It follows that there is a disk $D \subset H'$ that meets S only in a. Here a separates S into two disks, D_1, D_2. Set $S_1 = D \cup D_1$ and $S_2 = D \cup D_2$. Then both S_1 and S_2 are piecewise smooth 2-spheres in $\mathbb{R}^3$. We isotope S_1 and S_2 slightly (in a way that introduces only one new critical point) to be smooth. See Figures 3.14 and 3.15.

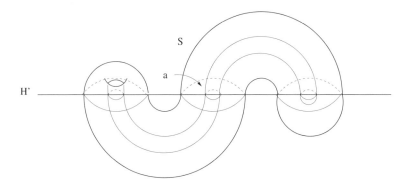

Figure 3.14. *A choice of a.*

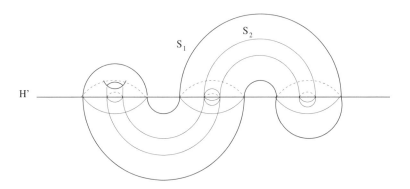

Figure 3.15. *The spheres S_1 and S_2.*

At least one of S_1, S_2, say S_2, has fewer saddles than S and hence bounds a 3-ball. Thus by Lemma 3.2.3, S is isotopic to S_1. Note that S_1 has no more saddles than S and, after a further isotopy as in Lemma 3.2.3, it has at least one fewer component of intersection with H'. (Note that the isotopies in Lemma 3.2.3 do not increase the number of saddles in S_1.) Thus, by induction on $(n, \#|H' \cap S|)$, S bounds a 3-ball. □

Corollary 3.2.6. $\mathbb{R}^3$, $\mathbb{B}^3$, *and* $\mathbb{S}^3$ *are irreducible.*

Proof. Let S be a 2-sphere submanifold in $\mathbb{R}^3$. The Schönflies Theorem tells us that S bounds a 3-ball. Thus $\mathbb{R}^3$ is irreducible. Let S' be a 2-sphere submanifold in the 3-ball. $\mathbb{B}^3$ is a subset of $\mathbb{R}^3$ that can be stretched (via an appropriate isotopy) to be arbitrarily large without stretching S'. The proof of the Schönflies Theorem takes place in a compact subset of $\mathbb{R}^3$ and thus can be adapted to take place in $\mathbb{B}^3$. Hence S' bounds a 3-ball in $\mathbb{B}^3$. Whence the 3-ball is irreducible. Finally, let S'' be a 2-sphere submanifold of $\mathbb{S}^3$. Here $\mathbb{S}^3 = \mathbb{R}^3 \cup \infty$. We may assume, after an isotopy, that S'' misses ∞; then $S'' \subset \mathbb{R}^3 \subset \mathbb{S}^3$. Thus the Schönflies Theorem furnishes a 3-ball bounded by S'' (in $\mathbb{R}^3 \subset \mathbb{S}^3$), whence $\mathbb{S}^3$ is irreducible. $\square$

We can say even more about $\mathbb{S}^3$: The 2-sphere S'' in $\mathbb{S}^3$ bounds a 3-ball to one side. If we remove a point (that we think of as ∞) from the 3-ball bounded by S'', we can repeat the argument in the proof of Corollary 3.2.6. This produces another 3-ball bounded by S''—to the other side!

The ideas used in the proof of the Schönflies Theorem carry over to other settings. We prove Alexander's Theorem via an argument analogous to that used in the proof of the Schönflies Theorem.

Theorem 3.2.7 (Alexander). *Let $\mathbb{T}$ be a torus submanifold in $\mathbb{S}^3$, then one of the components of $\mathbb{S}^3 \setminus \mathbb{T}$ has closure homeomorphic to a solid torus $\mathbb{S}^1 \times \mathbb{B}^2$.*

We will use the notion of "cobounding". The manifolds $L_1, \ldots, L_n$ are said to *cobound* the manifold $\tilde{L}$ if $\partial \tilde{L} = L_1 \cup \cdots \cup L_n$. For instance, below, we will see two simple closed curves in a surface that cobound an annulus.

Proof. Let $\mathbb{T} \subset \mathbb{R}^3$ be a torus submanifold. We isotope $\mathbb{T}$ so that the height function $h : \mathbb{R}^3 \to \mathbb{R}$ given by projection onto the third coordinate restricts to a Morse function $h|_\mathbb{T}$. By the Poincaré-Hopf Index Theorem,

$$\#\text{maxima} + \#\text{minima} - \#\text{saddles} = \chi(\mathbb{T}) = 0.$$

Denote the number of saddles by n. Note that since $\mathbb{T}$ is compact, it contains at least one maximum and one minimum. The Poincaré-Hopf Index Theorem then implies that $\mathbb{T}$ has at least two saddles.

If $n = 2$, then $\mathbb{T}$ has one maximum and one minimum. A level curve of $h|_\mathbb{T}$ near the maximum bounds a disk in the corresponding level surface of h. At the first saddle, this disk is either

(1) pinched into two disks (see Figure 3.16) or

(2) turns into a pinched annulus (see Figure 3.17).

Between the two saddles a level surface intersects $\mathbb{T}$ correspondingly either

(1) in two non-nested circles bounding disks or

(2) in two nested circles cobounding an annulus.

Below the second saddle (near the minimum), a level curve of $h|_\mathbb{T}$ again bounds a single disk. Thus at the second saddle, either

(1) the two non-nested circles bounding disks are wedged together or

(2) the annulus is pinched.

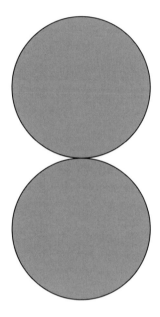

Figure 3.16. *A pinching into two disks.*

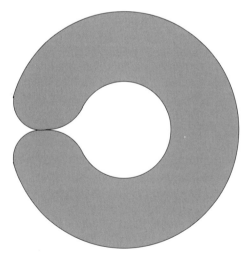

Figure 3.17. *Pinching into an annulus.*

3.2. The Schönflies Theorem

In both cases, these disks and annuli stack on top of each other to form a solid torus. See Figure 3.18 for an illustration of option (1). Note that in option (2), there is already a solid torus (annulus) $\times I$ between the two saddles.

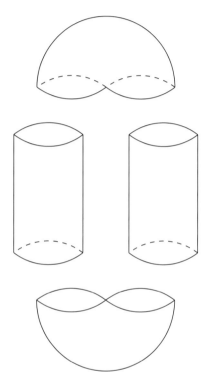

Figure 3.18. *Between the critical points.*

Suppose now that $n \geq 3$, and suppose that every torus in $\mathbb{R}^3$ with less than n saddles bounds a solid torus. Choose a regular level surface H, given by $z = r$, such that there are saddle points both above and below H and such that the number of components of $H \cap \mathbb{T}$ is minimal. We now argue by induction on $(n, \#|H \cap \mathbb{T}|)$.

Here $H \cap \mathbb{T}$ is a compact 1-manifold (without boundary), so $H \cap \mathbb{T}$ is a finite union of disjoint simple closed curves in H. Let a be an innermost component of $H \cap \mathbb{T}$ in H. It follows that there is a disk $D \subset H$ that meets S only in a. There are two cases:

Case 1. a is separating in $\mathbb{T}$.

Let P_1, P_2 be the surfaces obtained by cutting $\mathbb{T}$ along a. Since $\chi(P_i)$ is odd (an orientable surface with one boundary component), $\chi(P_i) \leq 2$, and $\chi(P_1) + \chi(P_2) = \chi(\mathbb{T}) = 0$, it must be the case that either P_1 or P_2, say

P_1, is a disk and that the other component, P_2, is a punctured torus. We proceed as in the proof of the Schönflies Theorem:

Set $S_1 = D \cup P_1$. Then S_1 is a piecewise smooth 2-sphere in $\mathbb{R}^3$ that can be isotoped to be a smooth 2-sphere. It thus bounds a 3-ball. By Lemma 3.2.3, $\mathbb{T}$ is isotopic to $(\mathbb{T}\backslash P_1) \cup D$. Now $(\mathbb{T}\backslash P_1) \cup D$ is a piecewise smooth torus with at most n saddles. It can be isotoped into a smooth torus with at most n saddles. By Lemma 3.2.3, $(\mathbb{T}\backslash P_1) \cup D$ can be isotoped to intersect H in one fewer level curve than $\mathbb{T} \cap H$. (Note that the isotopies in Lemma 3.2.3 do not increase the number of saddles of $(\mathbb{T}\backslash P_1) \cup D$.) Thus, by induction, $(\mathbb{T}\backslash P_1) \cup D$ and hence $\mathbb{T}$ bounds a solid torus.

Case 2. a is non-separating in $\mathbb{T}$.

Here $\mathbb{T}\backslash a$ is an annulus A. We may color the portion of this annulus that lies above H red and the portion that lies below H blue. Then every point in A is either colored red or colored blue or lies on one of the circles in $H \cap \mathbb{T}$. Just above a, A is colored red. Just below a, A is colored blue. Thus there must be at least one other component a' of $H \cap \mathbb{T}$ that is non-separating in $\mathbb{T}$. We may assume that a' is innermost in H. Indeed, if it is not, then it bounds a disk $\tilde{D}$ containing other components of $\mathbb{T} \cap H$; if an innermost component of $\mathbb{T} \cap H$ in $\tilde{D}$ is separating, proceed as in Case 1; if an innermost component of $\mathbb{T} \cap H$ in $\tilde{D}$ is non-separating, replace a' with this component.

Now $\mathbb{T}\backslash(a \cup a')$ consists of two annuli A_1, A_2. Denote the disk in $H\backslash\mathbb{T}$ bounded by a by D and that bounded by a' by D'. Set $S_1 = A_1 \cup D \cup D'$ and $S_2 = A_2 \cup D \cup D'$. Both S_1 and S_2 are piecewise smooth 2-spheres. Thus each of them bounds a 3-ball on either side. Two of these 3-balls can be identified along D and D' to form a solid torus. □

The following notion is related to irreducibility for manifolds with non-empty boundary.

Definition 3.2.8. A 3-manifold M is *boundary irreducible* if every simple closed curve c in ∂M that bounds a disk in M cuts M into two 3-manifolds one of which is a 3-ball.

Example 3.2.9. A boundaryless 3-manifold is necessarily boundary irreducible.

Example 3.2.10. The 3-ball is boundary irreducible.

The solid torus $V = \mathbb{S}^1 \times \mathbb{D}^2$ is not boundary irreducible. Indeed, consider the curve $m = \{p\} \times \partial \mathbb{D}^2 \subset \partial V$. Then m is non-separating in ∂V; hence the disk $D = \{p\} \times \mathbb{D}^2$ is also non-separating.

Definition 3.2.11. For V, m, D as above, m is called a *meridian* and D is called a *meridian disk*. A simple closed curve in ∂V that intersects m in one point is called a *longitude*.

Exercises

Exercise 1. Alexander's Theorem does not generalize to surfaces of genus greater than or equal to 2, in the sense that not every connected orientable 2-dimensional submanifold of $\mathbb{S}^3$ is the boundary of a regular neighborhood of a graph in $\mathbb{S}^3$. To convince yourself of this, draw a picture of a piecewise smooth closed orientable surface S of genus 2 in $\mathbb{S}^3$ so that neither component of $\mathbb{S}^3 \setminus S$ is a regular neighborhood of a graph in $\mathbb{S}^3$.

Exercise 2*. Let $\tilde{M}$ be a covering space of a 3-manifold M. Prove that $\tilde{M}$ is irreducible only if M is irreducible. (The converse is true but is more difficult to prove.)

Exercise 3. Use Exercise 2 to deduce that the 3-torus $\mathbb{T}^3$ is irreducible.

Exercise 4. Prove that the solid torus is irreducible.

3.3. 3-Manifolds that are Prime but Reducible

There are two compact connected 3-manifolds that deserve attention in the context of prime and reducible 3-manifolds. These two 3-manifolds are both $\mathbb{S}^2$-bundles over $\mathbb{S}^1$. In fact, they are both mapping tori with $S = \mathbb{S}^2$. One of these 3-manifolds is the product manifold $\mathbb{S}^2 \times \mathbb{S}^1$ ($= (\mathbb{S}^2 \times I)/\sim_{\mathrm{id}_{\mathbb{S}^2}}$). The other is also an (albeit twisted) mapping torus over $\mathbb{S}^2$.

Definition 3.3.1. We denote by $\mathbb{S}^2 \tilde{\times} \mathbb{S}^1$ the 3-manifold $(\mathbb{S}^2 \times I)/\sim_f$ where $f : \mathbb{S}^2 \to \mathbb{S}^2$ is the antipodal map. We also call $\mathbb{S}^2 \tilde{\times} \mathbb{S}^1$ the *twisted product* of $\mathbb{S}^2$ over $\mathbb{S}^1$.

We already discussed separating and non-separating simple closed curves in surfaces. The more general definition is analogous:

Definition 3.3.2. A submanifold A of a connected manifold X is *separating* if $X \setminus A$ has at least two components; otherwise it is *non-separating*.

Remark 3.3.3. Connected manifolds are path connected. It follows that a submanifold A of a connected manifold X is non-separating if and only if there is a simple closed curve in X that intersects A transversely in a single point. Moreover, it follows that a submanifold A of a connected manifold X is non-separating if and only if there is a simple closed curve in X that intersects A transversely in an odd number of points.

The 3-manifolds $\mathbb{S}^2 \times \mathbb{S}^1$ and $\mathbb{S}^2 \tilde{\times} \mathbb{S}^1$ contain non-separating 2-spheres. Theorem 3.3.4 below shows that this is a rare property. The proofs of many theorems in low-dimensional topology, and Theorem 3.3.4 is no exception, rely on the existence of regular neighborhoods. For more details see Appendix A.

Theorem 3.3.4. *An irreducible closed connected 3-manifold is prime. An orientable closed connected prime 3-manifold is either irreducible or $\mathbb{S}^2 \times \mathbb{S}^1$.*

Remark 3.3.5. More generally, a closed connected prime 3-manifold is either irreducible or $\mathbb{S}^2 \times \mathbb{S}^1$ or $\mathbb{S}^2 \tilde{\times} \mathbb{S}^1$, but we will not prove this fact here.

Proof. A non-trivial connected sum of 3-manifolds contains a sphere that does not bound a 3-ball. Hence an irreducible 3-manifold is prime.

Suppose M is prime and let S be a 2-sphere in M. If S is separating, then $M - S$ has two components, N_1, N_2. If neither N_1 nor N_2 is a 3-ball, then $M = N_1 \# N_2$ and M is not prime. Thus either N_1 or N_2 is a 3-ball; i.e., S bounds a 3-ball.

If S is non-separating, let a be a simple closed curve in M that intersects S once transversely. Furthermore, let $N(S)$ be a regular neighborhood of S in M and let $N(a)$ be a regular neighborhood of a. Both $N(a)$ and $N(S)$ are trivial I-bundles, so, in particular, $\partial N(S)$ consists of two copies of S. Transversality of S and a guarantees that $N(a)$ intersects $N(S)$ in a solid cylinder.

Let $\bar{M}$ be the submanifold of M obtained by removing the interior of $N(S) \cup N(a)$. Then $\partial \bar{M}$ is a 2-sphere $\bar{S}$ in M. Moreover, $\bar{S}$ is a separating 2-sphere. To one side of $\bar{S}$ is $N(S) \cup N(a)$. Hence there must be a 3-ball to the other side of $\bar{S}$. Therefore $M = N(S) \cup N(a) \cup (\text{3-ball}) = \mathbb{S}^2 \times \mathbb{S}^1$. □

In Section 5.5, we will prove that prime decompositions of 3-manifolds exist and that they satisfy an appropriately defined notion of uniqueness.

Exercises

Exercise 1. Construct a 2-to-1 covering map from $\mathbb{S}^2 \times \mathbb{S}^1$ to $\mathbb{S}^2 \tilde{\times} \mathbb{S}^1$.

Exercise 2. Consider the connected sum of $\mathbb{T}^3$ and $\mathbb{S}^2 \times \mathbb{S}^1$. How many non-isotopic 2-spheres decompose this manifold as a connected sum?

Exercise 3. Show that $\mathbb{S}^2 \tilde{\times} \mathbb{S}^1$ is non-orientable.

3.4. Incompressible Surfaces

In the study of surfaces, essential simple closed curves lying on the surface play an important role. They allow for the use of cut and paste techniques

3.4. Incompressible Surfaces

that suffice to classify surfaces. It is natural to try to proceed analogously in higher dimensions. This line of investigation has been pursued with a certain amount of success. We discuss the results in Chapter 5.

Some surfaces contained in a 3-manifold are more interesting than others. We here discuss one useful class of surfaces. This class of surfaces can be thought of as a generalization of the notion of "essential simple closed curve" in a 2-manifold.

Definition 3.4.1. A submanifold S in a compact n-manifold M is *proper* if $\partial S = S \cap \partial M$.

Convention 3.4.2. In what follows we will assume that a submanifold S of M is proper and write simply S (contained) in M or $S \subset M$, unless ∂S is explicitly said to lie elsewhere.

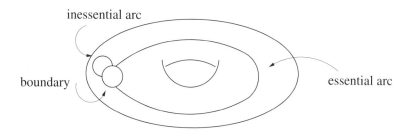

Figure 3.19. *An inessential arc and an essential arc in the punctured torus.*

Definition 3.4.3. A simple arc α in a surface F is *essential* if there is no simple arc β in ∂F such that $\alpha \cup \beta$ is a closed 1-manifold that bounds a disk in F. See Figure 3.19.

Definition 3.4.4. Let M be a 3-manifold. A surface S in M is *compressible* if either

- S is a 2-sphere bounding a 3-ball in M or
- there is a simple closed curve c in S that bounds a disk D with interior in $M \backslash S$ but that bounds no disk with interior a component of $S \backslash c$.

A surface that is not compressible is called *incompressible*. See Figures 3.20, 3.21, and 3.22. The disk D is called a *compressing disk* for S, or simply a compressing disk if the context is clear.

Example 3.4.5. If the surface $S \subset \mathbb{B}^3$ is incompressible, then S is a union of disks. Indeed, each component of ∂S is a simple closed curve in $\partial \mathbb{B}^3$. Such a curve bounds a disk to either side in $\partial \mathbb{B}^3$ and hence bounds a disk

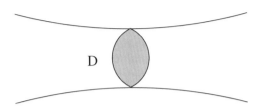

Figure 3.20. *A compressing disk for a surface.*

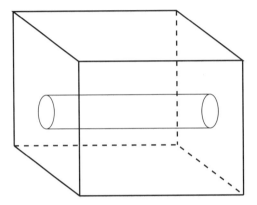

Figure 3.21. *A compressible 2-torus in the 3-torus.*

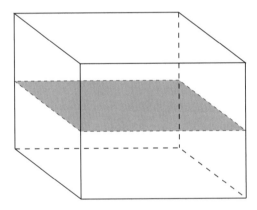

Figure 3.22. *An incompressible 2-torus in the 3-torus.*

in B. Thus each component of ∂S must bound a disk in S as well. You will show in the exercises that there are no closed incompressible surfaces in $\mathbb{B}^3$.

Definition 3.4.6. Let M be a 3-manifold. A surface $S \subset M$ is *boundary compressible*, or *∂-compressible*, if there is an essential simple arc α in S and an essential simple arc β in ∂M such that $\alpha \cup \beta$ is a closed 1-manifold that bounds a disk D in M with interior disjoint from S. See Figure 3.23. A surface that is not boundary compressible is *boundary incompressible*.

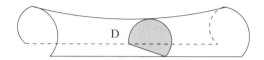

Figure 3.23. *A boundary compressing disk.*

Definition 3.4.7. Let M be a connected 3-manifold. A 2-sphere $S \subset M$ is *essential* if it does not bound a 3-ball. A surface $F \subset M$ is *boundary parallel* if it is separating and a component of $M \backslash F$ is homeomorphic to $F \times I$. A surface F in a 3-manifold M is *essential* if it is incompressible, boundary incompressible, and not boundary parallel.

The following definition honors Wolfgang Haken, who pioneered the study of incompressible surfaces in 3-manifolds via a machinery we will discuss in Chapter 5.

Definition 3.4.8. An orientable irreducible 3-manifold that contains a proper essential surface is called a *Haken 3-manifold*.

Remark 3.4.9. In subsequent sections and chapters we will always assume, unless stated otherwise, that submanifolds are proper.

Exercises

Exercise 1. Prove that the 2-torus in $\mathbb{T}^3$ pictured in Figure 3.22 is incompressible.

Exercise 2. Prove that a torus in $\mathbb{S}^3$ is necessarily compressible.

Exercise 3. Suppose that F is a surface and ϕ is a homeomorphism of F. Show that each copy $F \times \{\text{point}\}$ of F in the mapping torus of ϕ is incompressible.

Exercise 4*. Show that there are no closed incompressible surfaces in $\mathbb{B}^3$ or $\mathbb{S}^3$. (Hint: Adapt the proof of Alexander's Theorem.)

3.5. Dehn's Lemma*

Max Dehn was among the first to investigate knots with a view towards 3-manifolds. One of his aims was to answer a fundamental question in knot theory that we will encounter again in Chapter 4: how to decide whether or not a given simple closed curve in $\mathbb{S}^3$ bounds a disk. In 1910, in *Über die Topologie des dreidimensionalen Raumes* (see [**32**]), he believed he had (and was believed to have) established that this is the case if and only if the fundamental group of the complement of such a submanifold is abelian. But his proof hinged on a lemma he referred to as simply "the lemma". This lemma has since become known as "Dehn's Lemma".

Theorem 3.5.1 (Dehn's Lemma). *Suppose that M is a 3-manifold and $f : D \to M$ is a continuous map from the disk into M such that $f(\partial D) \subset \partial M$. If for some neighborhood U of ∂D, $f|_U$ is an embedding, then $f|_{\partial D}$ extends to an embedding.*

Dehn's original argument for his lemma was found to be incomplete, as pointed out by Kneser in 1929. In *Geschlossene Flächen in dreidimensionalen Mannigfaltigkeiten* (see [**80**]), Kneser believed he had extended Dehn's results. Then as his article went to press, he realized that Dehn's argument concerning the removal of double curves of an immersed disk was incomplete. He added a brief note to this effect at the end of his article [**80**, p. 260]. Dehn never fixed the proof of his lemma. It deserves to be mentioned that the teaching position he took on after fleeing Germany left him little time for research.

A proof of Dehn's Lemma was finally given by Christos Papakyriakopoulos in 1957; see [**119**]. In this proof, Papakyriakopoulos employed what he termed a "tower construction". Since then, this proof of Dehn's Lemma has undergone various revisions. For instance in Jaco and Rubinstein's *PL minimal surfaces in 3-manifolds* (see [**69**]), the authors introduce a PL theory of minimal surfaces and use it, among other things, to provide an alternate proof of Dehn's Lemma. In their proof, Papakyriakopoulos's universal covers are replaced by 2-fold covers. Here is a sketch of an argument:

Let $f : D \to M$ be a map as specified in Dehn's Lemma. It is then possible, though not obviously so, to assume that the self-intersections of $f(D)$ occur in the interior of $f(D)$ and are modeled on the two types indicated in Figures 3.24 and 3.25.

The "easy" case (the argument presented by Dehn covered this case): If the self-intersections of $f(D)$ occur along double curves and there are no triple points, then the preimage of such a double curve consists of two simple closed curves in D. Let a be an innermost simple closed curve in D that is the preimage of a double curve and let b be another simple closed curve such that $f(a) = f(b)$. Here a and b bound disks D_a and D_b in D. Two cases need to be considered: (1) D_a and D_b are disjoint; (2) $D_a \subset D_b$.

If D_a and D_b are disjoint, then we can alter f by swapping $f(D_a)$ and $f(D_b)$ (the image of D_a under f is replaced with $f(D_b)$ and vice versa; see Lemma 3.2.3). After isotoping the resulting map slightly, all assumptions of

3.5. Dehn's Lemma

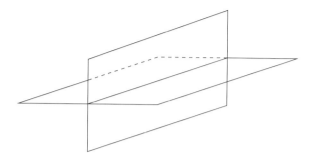

Figure 3.24. *A double curve.*

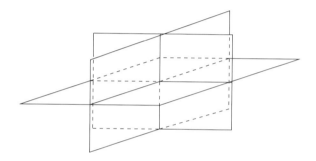

Figure 3.25. *A triple point.*

Dehn's Lemma are still met, but there are fewer curves of self-intersection in the image of the new map.

If $D_a \subset D_b$, then we can alter f by replacing $f(D_b)$ by $f(D_a)$. The result is a map satisfying all assumptions of Dehn's Lemma but whose image has fewer self-intersections. Thus the curves of self-intersection of $f(D)$ can be removed to produce the required embedding.

The "hard" case: If the self-intersections of $f(D)$ include triple points, we must proceed with more caution. Note that $f(D)$ is not a submanifold. Nevertheless, it is possible to construct an analog of a regular neighborhood of $f(D)$ in M. We denote this regular neighborhood by V_1. See Figure 3.26.

If V_1 has only 2-sphere boundary components, then $f(\partial D)$ (a simple closed curve on ∂V_1) bounds an embedded disk in M, as required. If V_1 has other surfaces as boundary components, then it has positive first Betti number. So it is the base space for many covering spaces. Let (M_1, V_1, p_1) be a 2-fold cover where M_1 is connected.

For each point in $f(D)$ there are exactly two points in $p_1^{-1}(f(D))$. Moreover, M_1 is homeomorphic to a regular neighborhood of $p_1^{-1}(f(D))$. Thus $p_1^{-1}(f(D))$ is connected. At the same time, $p_1^{-1}(f(D))$ is the lift of the image of the continuous map f. Since D is simply connected, f has two lifts to

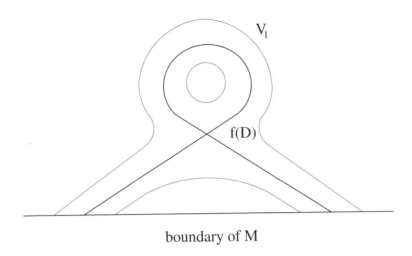

Figure 3.26. A "cross-section" of $f(D)$.

M_1. Thus $p_1^{-1}(f(D))$ is the image of a continuous map f_1 from the disjoint union of two disks into M_1. Here p_1 maps double curves and triple points of $p_1^{-1}(f(D))$ to double curves and triple points of $f(D)$, but not vice versa. Suffice it to say that a complexity argument allows the conclusion that there are fewer double curves in the image of at least one of these disks; call it D_1. Note that we can identify D with D_1 so that $(p \circ f_1)|_{D_1} = f$.

Now f_1, D_1, M_1 are as required in the hypotheses of Dehn's Lemma. Thus we can repeat the process above. We conclude that either $f_1(\partial D_1)$ bounds an embedded disk in M_1 or there is a connected 2-fold cover (M_2, V_2, p_2) with V_2 a regular neighborhood of $f_1(D_1)$ in M_1 and a continuous function $f_2 : D_2 \to M_2$ satisfying the hypotheses of Dehn's Lemma but such that $f_2(D_2)$ has fewer double curves than $f_1(D_1)$. If there are triple points in $f_2(D_2)$, we repeat the process. After some number of iterations, we reach a 2-fold cover (M_n, V_n, p_n) and a map $f_n : D_n \to M_n$ such that f_n is an embedding. (This alternating process of taking submanifolds and 2-fold covers is known as a "tower construction".)

Note that since f_n is an embedding, $p_n|_{f_n(D_n)}$ is at most 2-to-1. Thus since $(p_n \circ f_n)|_{D_n} = f_{n-1}$, $f_{n-1}(D_{n-1})$ has no triple points. Points at which $f_{n-1}(D_{n-1})$ is not embedded come from pairs of points of $f_n(D_n)$ and are hence double points. It follows from the argument in the "easy" case that $f_{n-1}(\partial D_{n-1})$ bounds an embedded disk in M_{n-1}. This gives an embedding of D_{n-1} into M_{n-1} that we continue to denote by $f_{n-1} : D_{n-1} \to M_{n-1}$. Then since $(p_{n-1} \circ f_{n-1})|_{D_{n-1}} = f_{n-2}$, $f_{n-2}(D_{n-2})$ has no triple points, etc. By induction, we can conclude that $f(\partial D)$ bounds an embedded disk in M.

3.5. Dehn's Lemma

The following quintuplet is attributed to John Milnor:

The perfidious Lemma of Dehn,

put many a good man to shame.

But Christos Pap-

akyriakop-

oulos proved it without any pain.

The tower construction also allowed Papakyriakopoulos to prove two other theorems that he called the Loop Theorem and the Sphere Theorem. Below is a generalization of the Loop Theorem formulated by John Stallings; see [**149**]:

Theorem 3.5.2 (The Loop Theorem). *Let M be a 3-manifold and F a connected surface in ∂M. If N is a normal subgroup of $\pi_1(F)$ and if $\ker(\pi_1(F) \to \pi_1(M))/N \neq \emptyset$, then there is a proper embedding $g : (D, \partial D) \to (M, F)$ such that $[g|_{\partial D}]$ is not in N.*

Corollary 3.5.3. *Let M be a 3-manifold and F a connected surface in ∂M. If $\ker(\pi_1(F) \to \pi_1(M)) \neq \{1\}$, then there is a proper embedding $g : (D, \partial D) \to (M, F)$ such that $[g|_{\partial D}]$ is non-trivial in $\pi_1(F)$.*

The geometric interpretation of the corollary above is that if the fundamental group of F does not inject into the fundamental group of M, then F is compressible.

Theorem 3.5.4 (The Sphere Theorem). *Let M be an orientable 3-manifold and N a $\pi_1(M)$-invariant subgroup of $\pi_2(M)$. If $\pi_2(M)/N \neq \emptyset$, then there is an embedding $g : \mathbb{S}^2 \to M$ such that $[g]$ is not in N.*

The common theme to Dehn's Lemma and the results mentioned above is that of promoting an "immersed" disk (a disk with double curves and triple points) to an embedded disk. In the setting of Dehn's Lemma this is achieved for a disk while fixing its boundary. In the setting of the Loop Theorem this is achieved for a disk while fixing the homotopy class of its boundary (not the boundary itself). In the setting of the Sphere Theorem this is achieved for a sphere. This common theme recurs quite frequently in the subject of 3-dimensional topology.

You will prove the following lemma in the exercises.

Lemma 3.5.5. *Let M be a 3-manifold. A surface $F \subset M$ is incompressible if and only if $\pi_1(F) \to \pi_1(M)$ is injective.*

Exercises

Exercise 1. The contrapositive of the Loop Theorem with $N = \langle 1 \rangle$ states that if a surface F in a 3-manifold M is incompressible, then $\pi_1(F) \to \pi_1(M)$ is injective. Show that the converse is true.

Exercise 2. Prove Lemma 3.5.5.

Exercise 3. Let F be a compact surface without boundary. List all, up to isotopy, incompressible boundary surfaces (those with and those without boundary) in $M = F \times I$.

3.6. Hierarchies*

The idea of a hierarchy parallels the process of cutting a surface along simple closed curves and simple arcs until only disks remain. Surfaces can be classified by reversing this procedure. This approach was not used in Chapter 2 but is discussed, for instance, in [**63**, Chapter 13]. See Figure 3.27. Codimension 1 submanifolds of a surface are curves and arcs. Codimension 1 submanifolds of a 3-manifold are surfaces. A hierarchy provides a sequence of surfaces that cuts a 3-manifold into 3-balls. The idea goes back to Wolfgang Haken, whose work we will discuss in Chapter 5.

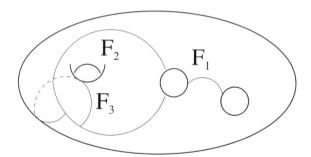

Figure 3.27. *The codimension 1 submanifolds in a 2-dimensional analog of a hierarchy.*

Definition 3.6.1. Let M be a compact 3-manifold. A *hierarchy* for M is a finite sequence of pairs $(M_1, F_1), \ldots, (M_n, F_n)$ such that

(1) F_i is a 2-sided incompressible and ∂-incompressible surface in M_i;

(2) $M_i = M_{i-1} \setminus \eta(F_{i-1})$, for $\eta(F_{i-1})$ an open regular neighborhood of F_{i-1} (we also say that M_i is obtained by *cutting* M_{i-1} along F_{i-1});

(3) $M_1 = M$;

(4) each component of M_{n+1} is a 3-ball.

3.6. Hierarchies

As an example, we illustrate a hierarchy for the 3-torus: Here (M_1, F_1) consists of the 3-torus M_1 and an incompressible torus F_1 contained in M_1. See Figure 3.28. The second pair, (M_2, F_2), consists of torus $\times I$ and an incompressible annulus therein. See Figure 3.29. The third pair, (M_3, F_3), consists of annulus $\times I$ and an essential disk therein. See Figure 3.30. Cutting M_3 along F_3 yields a 3-ball. So this is a hierarchy.

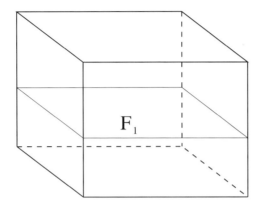

Figure 3.28. *An incompressible torus in the 3-torus.*

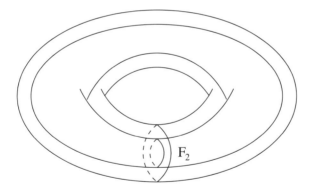

Figure 3.29. *An incompressible annulus in torus $\times I$.*

Recall that an orientable irreducible 3-manifold is Haken if it contains a 2-sided incompressible surface. Haken 3-manifolds played a major role in the study of 3-dimensional topology in the 1960s, 1970s, and 1980s. We are restricting our attention to the case of orientable 3-manifolds, though many of the results discussed here hold, in a more general form, for non-orientable 3-manifolds. The main theorem concerning hierarchies is the following:

Theorem 3.6.2. *A compact orientable irreducible 3-manifold M is Haken if and only if M has a hierarchy.*

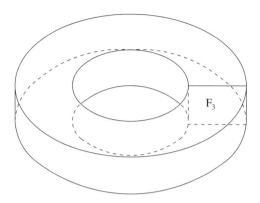

Figure 3.30. *A disk in annulus × I.*

Theorem 3.6.2 tells us that a given Haken manifold can be constructed by taking an appropriate collection of 3-balls and identifying appropriate portions of their boundaries.

A key ingredient in proving this theorem is the following lemma:

Lemma 3.6.3. *Suppose M is a compact orientable 3-manifold such that ∂M contains a surface of positive genus. Then M contains a properly embedded, 2-sided, incompressible, ∂-incompressible surface F such that $\{1\} \neq [\partial F] \in \pi_1(\partial M)$.*

The proof of this lemma is not hard but relies on an understanding of homology. It can be found in [**67**, Theorem III.10] or [**63**, Lemma 6.8].

Definition 3.6.4. The *length* of the hierarchy $(M_1, F_1), \ldots, (M_n, F_n)$ is n.

Haken manifolds admit hierarchies of a very special type:

Theorem 3.6.5. *Let M be a Haken manifold. Then M has a hierarchy of length 4.*

Specifically, a Haken manifold admits a hierarchy
$$(M_1, F_1), (M_2, F_2), (M_3, F_3), (M_4, F_4)$$
where F_1 consists of disjoint closed surfaces in M, F_2 consists of disjoint surfaces with non-empty boundary in M_2, F_3 consists of disjoint annuli, and F_4 consists of disjoint disks.

This is [**67**, Theorem IV.19].

Remark 3.6.6. In Theorem 3.6.5, F_4 consists of disks and M_5 consists of 3-balls. It follows that M_4 is a disjoint union of handlebodies.

The existence of the hierarchy of length 4 provided by Theorem 3.6.5 is proved by employing important existence and maximality results. For

3.6. Hierarchies

instance, F_1 is a maximal collection of disjoint non-parallel closed incompressible surfaces in M_1. If M_1 is closed, then a collection of disjoint non-parallel closed incompressible surfaces in M_1 exists because M_1 is Haken. The fact that a finite maximal such collection exists follows from Theorems 5.4.3 and 5.4.5 (together known as Kneser-Haken finiteness). The collections of surfaces F_2, F_3 exist by Lemma 3.6.3, and Kneser-Haken finiteness again establishes the fact that there are finite maximal collections of disjoint non-parallel incompressible surfaces of the specified type.

The cut and paste techniques that provide the basis for understanding 2-manifolds have a chance of succeeding in the context of Haken manifolds. This can be seen as a motivating principle behind some of Haken's work and also for the theorem of Waldhausen mentioned below. However, the extra dimension does add difficulties. In addition, note that many (some would say "most") 3-manifolds are not Haken.

Definition 3.6.7. Two topological spaces X, Y are *homotopy equivalent* if there are continuous maps $f : X \to Y$ and $g : Y \to X$ such that $g \circ f$ is homotopic to id_X and $f \circ g$ is homotopic to id_Y.

A key result on Haken manifolds is due to Friedhelm Waldhausen; see [156].

Theorem 3.6.8 (Waldhausen's Theorem). *Homotopy equivalent closed Haken manifolds are homeomorphic.*

This theorem does not hold for 3-manifolds in general. Consider for instance the following construction: Let T_1, T_2 be solid tori. Here $T_i = \mathbb{B}^2 \times \mathbb{S}^1$. We denote the simple closed curve $(\partial \mathbb{B}^2) \times \{\text{point}\}$ by m_i and we denote the simple closed curve $p \times \mathbb{S}^1$, where $p \in \partial \mathbb{B}^2$, by l_i. Now identify T_1 and T_2 along their boundaries via the element $[h]$ of the mapping class group of ∂T_i (see Lemma 2.6.2) corresponding to the matrix

$$A_h = \begin{bmatrix} p & r \\ q & s \end{bmatrix}$$

where p, q are relatively prime integers and r, s are such that $ps + qr = 1$. (Thus m_1 is identified with a simple closed curve in ∂T_2 that meets m_2 in q (equivalently oriented) points and l_2 in p (equivalently oriented) points.)

While Haken manifolds constitute what might look like a small subset of 3-manifolds, recent work of Agol shows that if M is an irreducible 3-manifold, then M is finitely covered either by $\mathbb{S}^3$ or by a Haken manifold. See [2].

Definition 3.6.9. A 3-manifold that is constructed as above is called a *lens space*. It is denoted by $L(p, q)$.

As it turns out, $L(7,1)$ and $L(7,2)$ are homotopy equivalent but not homeomorphic. (See [**63**, Lemma 3.23] and [**127**, p. 235] and the exercises.)

Theorem 3.6.10 (The Poincaré Conjecture). *If M is a closed 3-manifold that is homotopy equivalent to $\mathbb{S}^3$, then M is homeomorphic to $\mathbb{S}^3$.*

The Poincaré Conjecture was an open conjecture for over a century. It was finally proved by G. Perelman in the early 21st century using geometric and analytic methods.

We now describe the Whitehead manifold. It is another manifold that is of interest in the context of Waldhausen's Theorem and the Poincaré Conjecture. The Whitehead manifold is an example of an irreducible 3-manifold that is open, connected, contractible, and hence homotopy equivalent to the open 3-ball, yet it is not homeomorphic to the open 3-ball. We construct the Whitehead manifold as follows: Consider the solid torus $T \subset \mathbb{S}^3$ as in Figure 3.31. Map T into itself via the map $h : T \to T$ so that $h(T)$ lies inside of T as in Figure 3.31.

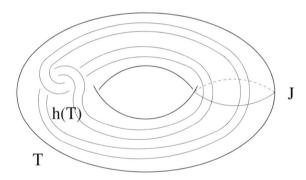

Figure 3.31. *The solid torus $h(T)$ within the solid torus T.*

Now consider the solid torus $h(T)$ and map $h(T)$ into itself in the same way to obtain $h^2(T)$. Continuing in this fashion yields an infinite sequence of nested solid tori T, $h(T)$, $h^2(T)$, ... Set $X = \bigcap h^n(T)$. The Whitehead manifold is $W = \mathbb{S}^3 \backslash X$.

Proposition 3.6.11. *The Whitehead manifold is open, irreducible, connected, and simply connected.*

Proof. Here we set
$$W_0 = \mathbb{S}^3 \backslash T$$
and
$$W_n = \mathbb{S}^3 \backslash X_n.$$

3.6. Hierarchies

Then W_n is open,
$$W_n = W_0 \cup W_1 \cup W_2 \cup \cdots \cup W_n,$$
and
$$W = W_0 \cup W_1 \cup W_2 \cup \ldots.$$

In particular, W is open. Note that in $\mathbb{S}^3$, $h(T)$ is isotopic to T; see Figure 3.32. (The isotopy is accomplished by "untwisting" the "inner" portion of $h(T)$ as in Figure 3.32 and then shortening the same portion.) Thus, $\forall n$ W_n is homeomorphic to W_0.

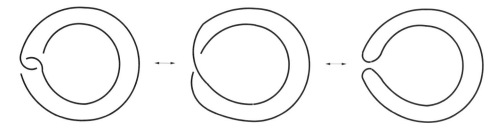

Figure 3.32. *The solid torus $h(T)$ is isotopic to T.*

To see that W is irreducible, consider a 2-sphere submanifold S in W. Since S is compact and $\{W_n\}$ is an open cover for W, $\exists n$ such that $S \subset W_n$. Now $W_n \subset \mathbb{S}^3$, so $S \subset \mathbb{S}^3$ where it is separating and bounds 3-balls to either side. Since X_n is connected, it lies in only one of these 3-balls; thus the other 3-ball bounded by S lies in W_n and hence in W.

To see that W is connected, note that a manifold is connected if and only if it is path-connected. So suppose that $x, y \in W$. By the same reasoning as above, $\exists m$ such that $x, y \in W_m$. Since W_m is path-connected, there is a path from x to y lying in W_m and hence in W.

To see that W is simply connected, consider an embedded curve c in W. Since c is compact, by the same reasoning as above, $\exists l$ such that $c \subset W_l$. It hence suffices to show that $\pi_1(W_l)$ is trivial in $\pi_1(W)$. We will show that $\pi_1(W_l)$ is trivial in $\pi_1(W_{l+1})$ by establishing that $\pi_1(W_0)$ is trivial in $\pi_1(W_1)$.

Consider the curve J in Figure 3.31. Here J is the generator for $\pi_1(W_0)$. The isotopy in Figure 3.32 affects J as pictured in Figure 3.33. After a homotopy, J is as in Figure 3.34 and thus bounds a disk in W_1. $\square$

Remark 3.6.12. In fact, the Whitehead manifold is contractible. This is a non-trivial consequence of the fact that it is irreducible and simply connected: By the Sphere Theorem, $\pi_2(W) = \{1\}$ since W is irreducible. It follows from a theorem of Whitehead that a connected open 3-manifold

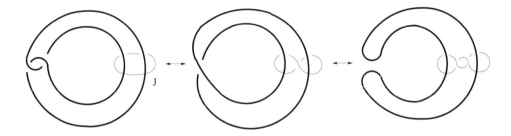

Figure 3.33. *The curve J.*

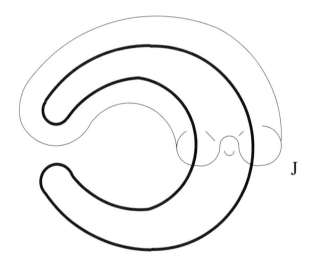

Figure 3.34. *The curve J.*

with trivial fundamental group and second homotopy group is homotopic to a point.

The following definition is of interest in this context. We will not use it otherwise.

Definition 3.6.13. A topological space W is said to be *simply connected at infinity* if for every compact subset $K \subset W$ there is a compact set K' such that $K \subset K' \subset W$ and the inclusion map $W \setminus K' \hookrightarrow W \setminus K$ induces the trivial map on fundamental groups.

Theorem 3.6.14 (Brin-Thickstun)**.** *An open, connected, contractible 3-manifold that is simply connected at infinity is homeomorphic to the open 3-ball.*

Exercises

Exercise 1*. Prove that $L(7,1)$ and $L(7,2)$ are homotopy equivalent but not homeomorphic. (This is challenging. See [**63**, Lemma 3.23] for a proof that $L(7,1)$ and $L(7,2)$ are homotopy equivalent. To see that they are not homeomorphic, see the discussion in [**127**, p. 235] and the references therein.)

Exercise 2. Prove that the Whitehead manifold is not simply connected at infinity.

Exercise 3. Prove that the Whitehead manifold is not homeomorphic to the open 3-ball.

3.7. Seifert Fibered Spaces

Seifert fibered spaces constitute an important class of 3-manifolds. For details, see [**146**]. Not all of them are Haken, but many are. We discuss them here to provide more examples of 3-manifolds in general and of Haken manifolds in particular. Much can be said about Seifert fibered spaces. We will only touch on some of the highlights. One of the reasons for the continued interest in Seifert fibered spaces is the fact that they project in a natural way onto a 2-dimensional "base space". This fact makes them particularly amenable to computations.

Definition 3.7.1. The *fibered solid torus* of type (l,m) is a solid torus partitioned into circles obtained as follows: Consider the cylinder $\mathbb{D}^2 \times [-1,1]$. It is partitioned into intervals of the form $\{\text{point}\} \times [-1,1]$. Identify $\mathbb{D}^2 \times \{-1\}$ to $\mathbb{D}^2 \times \{1\}$ by setting $(re^{i2\pi\theta}, 1)$ equal to $(re^{i2\pi(\theta+\frac{m}{l})}, -1)$, for $m \in \mathbb{N} \cup \{0\}$ and $l \in \mathbb{N}$. A circle formed by intervals of the form $\{\text{point}\} \times [-1,1]$ is called a *fiber* of the fibered solid torus.

The *core* of a fibered solid torus is the fiber obtained from $\{0\} \times I$. If $l > 1$, then we call the fibered solid torus T of type (l,m) an *exceptionally fibered* solid torus. In this case the core of T is called an *exceptional fiber* and all other fibers of T are called *regular fibers*. If $l = 1$, then we call a fibered solid torus T of type (l,m) a *regularly fibered* solid torus. In this case all of the fibers of T are called regular fibers. See Figure 3.35. In addition, when $l = 1$, the disk $D = \mathbb{D}^2 \times \{0\} \subset T$ intersects each fiber once, so there is a bundle $(T, \mathbb{S}^1, D, p)$, where $p : T \to D$ send each fiber to its point of intersection with D.

Suppose T_1 and T_2 are fibered solid tori. A *fiber-preserving* homeomorphism between T_1 and T_2 is a homeomorphism $h : T_1 \to T_2$ that takes fibers to fibers.

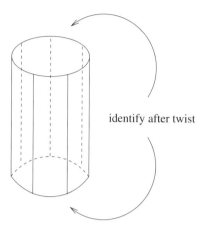

Figure 3.35. *A fibered solid torus.*

Remark 3.7.2. We may assume that $0 \leq m < \frac{l}{2}$. Indeed, any fibered solid torus is homeomorphic via a fiber-preserving homeomorphism to a fibered solid torus of type (l, m) with l, m satisfying this constraint. With this assumption, the existence of a fiber-preserving homeomorphism between a fibered solid torus of type (l_1, m_1) and a fibered solid torus of type (l_2, m_2) necessitates $l_1 = l_2$ and $m_1 = m_2$.

Definition 3.7.3. A *fibered neighborhood* of a simple closed curve f in a 3-manifold M that consists of simple closed curves is

- a closed regular neighborhood homeomorphic to a fibered solid torus via a fiber-preserving homeomorphism when f is in the interior of M or
- a closed regular neighborhood homeomorphic to a product (half-disk) $\times \, \mathbb{S}^1$ partitioned into fibers $\{\text{point}\} \times \mathbb{S}^1$ when $f \subset \partial M$.

Definition 3.7.4. An orientable 3-manifold M is a *Seifert fibered space* if M is the union of pairwise disjoint simple closed curves called *fibers* such that each fiber in M has a closed regular neighborhood that is a fibered neighborhood. The description of M in terms of fibers is called the *Seifert fibration* of M.

Denote the quotient space of M obtained by identifying each fiber to a point by B and denote the corresponding quotient map by $p : M \to B$. Here B is called the *base space* of M.

Analogous to fibered solid tori, there are also fibered solid Klein bottles. These allow an extension of the notion of Seifert fibered space to non-orientable 3-manifolds, but we will not pursue this more general topic here.

3.7. Seifert Fibered Spaces

Remark 3.7.5. Note that an $\mathbb{S}^1$-bundle over a surface is a Seifert fibered space (with no exceptional fibers, whether or not the bundle is trivial). However, a Seifert fibered space is not necessarily an $\mathbb{S}^1$-bundle over a surface. (The complement of the exceptional fibers of a Seifert fibered space is an $\mathbb{S}^1$-bundle over a surface.)

Lens spaces are examples of Seifert fibered spaces. To see the fibers of $L(p, q)$, recall that $L(p, q)$ is constructed by identifying two solid tori T_1, T_2 along their boundaries via a homeomorphism corresponding to

$$A_h = \begin{bmatrix} p & r \\ q & s \end{bmatrix}.$$

Each of the two solid tori is a fibered solid torus in many different ways. We need to describe T_1, T_2 as fibered solid tori in such a way that the fibers match up along $\partial T_1 = \partial T_2$. There are several ways to do this. One possibility is the following: Using the notation in the discussion preceding Definition 3.6.9, choose a simple closed curve c on $\partial T_1 = \partial T_2$ that intersects m_1 once. Any simple closed curve that intersects m_2 in $r \neq 0$ (equivalently oriented) points and l_2 in s (equivalently oriented) points will do. Now there is exactly one way in which T_i is a fibered solid torus with c as a fiber. This decomposes $L(p, q)$ into fibers as required.

Remark 3.7.6. The above examples illustrate the lack of uniqueness of a Seifert fibration. There are infinitely many ways to fiber a solid torus. This translates into infinitely many different Seifert fibrations for a lens space. Interestingly, Seifert fibrations for more complicated Seifert fibered spaces are unique up to isotopy. See Theorem 3.7.19 below.

The 3-sphere arises as a lens space, specifically, $\mathbb{S}^3 = L(1, n)$. The exceptional fibers of the Seifert fibration of $\mathbb{S}^3$ are pictured in Figure 3.36.

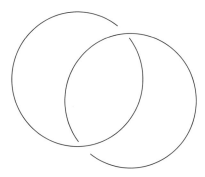

Figure 3.36. *The exceptional fibers of a Seifert fibration of $\mathbb{S}^3$.*

Example 3.7.7. Recall the non-trivial I-bundle over the Möbius band described in Example 3.1.10. The total space of this bundle is a solid torus, which can be viewed as a fibered solid torus in many different ways and is hence a Seifert fibered space.

Analogously, we can define a non-trivial I-bundle over the Klein bottle, called the *twisted I-bundle* over the Klein bottle. The total space of this bundle can be constructed from a cube by appropriately identifying the sides (front to back and left to right). The cube consists of intervals running front to back and also consists of intervals running left to right. In both cases, the intervals yield fibers of a Seifert fibration for the resulting total space. Thus the twisted I-bundle over the Klein bottle is a Seifert fibered space and it admits distinct Seifert fibrations.

Definition 3.7.8. Let M be a 3-manifold with non-empty boundary. The *double* of M is the 3-manifold obtained from two copies of M that have been identified (via the identity map) along their boundaries.

Example 3.7.9. The double of a Seifert fibered space is also a Seifert fibered space.

Lemma 3.7.10. *If T_1 and T_2 are fibered neighborhoods of the fiber f, then f is an exceptional fiber of T_1 if and only if f is an exceptional fiber of T_2.*

Proof. This is a consequence of the uniqueness of regular neighborhoods; see Theorem 3.1.13 and Remark 3.7.2. □

Lemma 3.7.10 allows us to refer to the fibers of a Seifert fibered space as exceptional fibers or regular fibers unambiguously.

Remark 3.7.11. The base space B of a Seifert fibered space is a surface. This can be seen as follows: The quotient of a fibered solid torus is a disk. Let q be a point in B. Then $p^{-1}(q)$ is a fiber f of M. The fibered neighborhood T of f hence yields a neighborhood $p(T)$ of q that is homeomorphic to a disk. If f is an exceptional fiber, then $p(f)$ is called an *exceptional* point (also called a *cone* point). Sometimes we attach a number to an exceptional point. This number represents the number l of the (exceptionally) fibered solid torus neighborhood of f. Thus we think of B as a surface with exceptional points.

We can describe several Seifert fibered spaces in terms of their base space. One class of Seifert fibered spaces is called prism manifolds. Prism manifolds are Seifert fibered spaces with base space a 2-sphere and three exceptional fibers. Some prism manifolds, albeit a minority, are Haken manifolds. For more on Seifert fibered spaces and Haken manifolds see Theorem 3.7.18.

3.7. Seifert Fibered Spaces

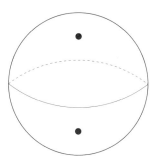

Figure 3.37. *The base space of a lens space.*

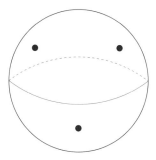

Figure 3.38. *The base space of a prism manifold.*

Remark 3.7.12. The interiors of the fibered neighborhoods of fibers form an open cover for a Seifert fibered space. It follows that in a compact Seifert fibered space, there will be only finitely many exceptional fibers.

Definition 3.7.13. A subset Z of X is *saturated* with respect to $p: X \to Y$ if $Z = p^{-1}(p(Z))$.

In the construction of lens spaces, two solid tori were identified along their boundaries. The torus resulting from the identification of the boundary tori is a saturated torus with respect to the quotient map. It corresponds to a circle in the base space. See Figure 3.37.

The Seifert fibration of a Seifert fibered space provides a means of describing and isotoping surfaces. Incompressible surfaces in Seifert fibered spaces can be described completely. (See Theorem 3.7.16.) As a warm-up, we will prove a lemma.

Lemma 3.7.14. *A properly embedded connected orientable incompressible surface F in a solid torus V is either a disk or an annulus.*

Proof. We first show that a solid torus V is irreducible. Here V is homeomorphic to the solid torus
$$\{(x,y,z) \in \mathbb{R}^3 : (\sqrt{x^2+y^2}-2)^2 + z^2 = 1\}.$$
Thus we may assume that $V \subset \mathbb{R}^3 \subset \mathbb{S}^3$. Now given a 2-sphere $S \in V$, $S \in \mathbb{S}^3$, where it bounds 3-balls to either side. Since ∂V is connected and disjoint from S, one of these 3-balls lies entirely in V. Hence V is irreducible.

Let V be a solid torus and suppose that $F \subset V$ is incompressible. Let D be a meridian disk of V. The proof is by induction on the number of components of $F \cap D$. If $F \cap D$ is empty, then F is an incompressible surface in the 3-ball $V \backslash D$ and is hence a disk. (See Example 3.4.5.) Simple closed curves in $F \cap D$ can be eliminated via a standard innermost disk argument; see Lemma 3.2.3 and Example 3.9.1. Thus $F \cap D$ consists of arcs.

For C a component of ∂F, either C is inessential or essential in ∂V. In the case of the former, C bounds a disk in $\partial V \subset V$. Since F is incompressible, this implies that F is a disk. Thus we may assume, in what follows, that all components of ∂F are essential in ∂V. It follows that the components of ∂F are parallel essential simple closed curves in ∂V. We isotope F to minimize the number of components of $F \cap D$.

Suppose that $F \cap D \neq \emptyset$. Let a be an outermost arc in D. You will show in the exercises that the endpoints of a lie on distinct components of ∂F. Let b, c be the two components of ∂F containing the endpoints of a. Let $N(a \cup b \cup c)$ be a regular neighborhood of $a \cup b \cup c$ in F. It is a thrice-punctured sphere. You will show in the exercises that the boundary component of $N(a \cup b \cup c)$ that lies in the interior of F bounds a disk in M. It hence bounds a disk in F. Thus $F = N(a \cup b \cup c) \cup (\text{disk})$ is an annulus. $\square$

To further understand surfaces lying in Seifert fibered spaces, the following definitions prove useful:

Definition 3.7.15. A surface in a Seifert fibered space is *vertical* if it saturated, i.e., if it consists of fibers. A surface in a Seifert fibered space is *horizontal* if it is everywhere transverse to the fibration.

Saturated annuli and tori are examples of vertical surfaces in Seifert fibered spaces. Covering spaces of the base space of a Seifert fibered space can sometimes be embedded in the Seifert fibered space. (This fact accounts for those prism manifolds that are Haken.) When this is possible, such a surface is an example of a horizontal surface in a Seifert fibered space.

Theorem 3.7.16. *An orientable essential surface in an orientable Seifert fibered space can be isotoped so that it is either horizontal or vertical.*

3.7. Seifert Fibered Spaces

The proof of this theorem is lengthy; see [**67**, Theorem VI.34].

We conclude this section with a few well-known theorems that provide a brief overview of further results on Seifert fibered spaces.

Theorem 3.7.17. *An orientable Seifert fibered space is either irreducible or homeomorphic to $\mathbb{S}^2 \times \mathbb{S}^1$ or $\mathbb{RP}^3 \# \mathbb{RP}^3$.*

Proof. Let M be an orientable Seifert fibered space. Two cases need to be considered:

Case 1. $\partial M \neq \emptyset$.

Let B be the base space of M. Then $\partial B \neq \emptyset$. The proof is by induction on the pair $(-\chi(B), \# \text{exceptional points})$ (in the dictionary order). Suppose first that B is a disk with at most one exceptional point. Then M is a solid torus. The first paragraph in the proof of Lemma 3.7.14 established that a solid torus is irreducible.

If there are at least two exceptional points in B, then there is an arc a in B that cuts off a disk containing exactly one exceptional point. The arc a corresponds to an essential saturated annulus $p^{-1}(a)$. Similarly, if $-\chi(B) \geq 0$, then there is an essential arc in B corresponding to an essential saturated annulus in M. Denote the essential saturated annulus by A.

Let S be a 2-sphere in M. If S is disjoint from A, then S lies in $M \backslash A$, the interior of a Seifert fibered space with boundary. By the inductive hypothesis, this Seifert fibered space is irreducible. Hence S bounds a 3-ball in $M \backslash A$. Thus S bounds a 3-ball in M.

Suppose that $A \cap S \neq \emptyset$. Let c be a component of $A \cap S$ that is innermost in S. Then one component of $S \backslash c$ is a disk D that is disjoint from S. In the exercises, you will prove a result that implies that A is incompressible. Since A is incompressible, c is inessential in A and hence one component, D', of $A \backslash c$ is a disk. Here $D \cup D' \cup c$ is a 2-sphere that, after a small isotopy, lies in one component of $M \backslash A$ and hence bounds a 3-ball. Proceeding as in Lemma 3.2.3 provides an isotopy of S reducing the number of components in $A \cap S$. Thus if we isotope S so that the number of components of $A \cap S$ is minimal, then $A \cap S = \emptyset$. Above we proved that if S is disjoint from A, then S bounds a 3-ball.

Case 2. $\partial M = \emptyset$.

If M contains a saturated incompressible torus T, then the argument above can be applied to show that a 2-sphere $S \subset M$ can be isotoped to be disjoint from T. Thus S lies in $M \backslash T$, the interior of a Seifert fibered space with boundary. It follows that S bounds a 3-ball in $M \backslash T$ and hence in M.

It remains to consider the cases where B is either a 2-sphere with at most three exceptional fibers or a projective plane with at most one exceptional fiber. In the exercises you will prove that in all other cases M contains a saturated incompressible torus.

The case of lens spaces ($B = \mathbb{S}^2$ with up to two exceptional points) is left as an exercise. As it turns out, a lens space is either irreducible or homeomorphic to $\mathbb{S}^2 \times \mathbb{S}^1$. In addition, in the case of prism manifolds ($B = \mathbb{S}^2$ with three exceptional points) we refer to a theorem of Waldhausen, which states that prism manifolds are irreducible.

To treat the remaining case, suppose that B is a projective plane with at most one exceptional point. Let t be a simple closed curve in B such that one component of $B \setminus t$ is a disk containing the exceptional point if there is one. Let $T = p^{-1}(t)$ be the corresponding saturated torus. Since t is separating, so is T. T cuts M into two components: (1) M', a circle bundle over the Möbius band with bundle structure coming from the Seifert fibration and (2) V, a fibered solid torus. Since M, and thus M', is orientable, M' must be the twisted circle bundle over the Möbius band.

Suppose S is a 2-sphere in M. If S is disjoint from T, then it lies in either M' or V and hence bounds a 3-ball. We may thus assume, in what follows, that $S \cap T$ is not empty. Let c be a component of $S \cap T$ that is innermost in S. Then at least one component of $S \setminus c$ is a disk D that is disjoint from T. If c is inessential in T, then a component of $T \setminus c$ is a disk D' and $D \cup D' \cup c$ is a 2-sphere that, after isotopy, lies in a Seifert fibered space with boundary where it bounds a 3-ball. By Lemma 3.2.3, there is an isotopy reducing the number of components of $S \cap T$. Thus, if the number of components of $S \cap T$ is minimal, then $S \cap T$ consists of curves essential in T.

We make the following observations (the proofs are left to the reader): We assume that the number of components of $S \cap T$ is minimal. Under this assumption:

(1) there are no components of $S \cap M'$ or $S \cap V$ that are boundary compressible via a boundary compressing disk that meets more than one component of $S \cap T$;

(2) $\partial M'$ is incompressible in M';

(3) a properly embedded orientable incompressible surface with non-empty boundary in M' that is not boundary compressible via a boundary compressing disk that meets more than one component of $S \cap T$ must be an annulus that double covers the Möbius band.

It now follows that S must consist of an annulus in M' that double covers the Möbius band together with two disks in V. Thus S cuts M into two identical pieces, each consisting of one of the two twisted I-bundles over a Möbius band coming from M' and one of the 3-balls coming from V. Each of these pieces is homeomorphic to a once-punctured $\mathbb{RP}^3$. □

Theorem 3.7.18. *A closed orientable Seifert fibered space M is either a Haken manifold, a lens space (including $\mathbb{S}^2 \times \mathbb{S}^1$, $\mathbb{S}^3$), $\mathbb{RP}^3 \# \mathbb{RP}^3$, or a prism manifold. In the latter case, M is Haken if and only if $H_1(M)$ is infinite.*

See [67, Theorem VI.15].

Theorem 3.7.19. *The only Seifert fibered spaces with non-unique fiberings are:*

(a) *lens spaces (including $\mathbb{S}^2 \times \mathbb{S}^1$, $\mathbb{S}^3$);*

(b) *prism manifolds (only in some cases);*

(c) *the solid torus;*

(d) *the twisted I-bundle over the Klein bottle;*

(e) *the double of the twisted I-bundle over the Klein bottle (this 3-manifold also fibers over $\mathbb{S}^2$ with four exceptional fibers for which $l_1 = l_2 = l_3 = l_4 = 2$).*

See [67, VI.16].

The following theorem was long known as the Topological Seifert Conjecture. It was finally proved by a series of results due to Tukia, Mess, Gabai, Casson-Jungreis, and Scott.

Theorem 3.7.20. *Let M be a compact orientable irreducible 3-manifold with $\pi_1(M)$ infinite. Then M is a Seifert fibered space if and only if $\pi_1(M)$ has a normal subgroup isomorphic to $\mathbb{Z}$.*

Exercises

Exercise 1. Prove Remark 3.7.2.

Exercise 2. Let V, D, a, F be as in Lemma 3.7.14 and assume that F has been isotoped so that the number of components of $F \cap D$ is minimal. Show that the endpoints of a lie on distinct components of ∂F.

Exercise 3. Let $V, D, a, F, b, c, N(a \cup b \cup c)$ be as in Lemma 3.7.14 and assume that F has been isotoped so that the number of components of $F \cap D$ is minimal. Observe that a cuts off a disk $\hat{D}$ from D. Here $b \cup c$ cuts ∂V into two annuli. One of these annuli, call it A, meets $\hat{D}$. Use A together with $\hat{D}$ to show that the component $\partial N(a \cup b \cup c) \setminus (b \cup c)$ bounds a disk in M.

Exercise 4. Prove the observations near the end of the proof of Theorem 3.7.17.

Exercise 5. Prove that lens spaces are irreducible.

Exercise 6. Prove that a saturated annulus in a Seifert fibered space that cuts off an exceptionally fibered solid torus to one side and a Seifert fibered space that is not fibered over a disk or that has at least two exceptional fibers to the other side is incompressible.

Exercise 7. What hypotheses guarantee that a saturated annulus or torus is incompressible?

3.8. JSJ Decompositions

JSJ decompositions provide a geometric decomposition of 3-manifolds. The acronym stands for Jaco, Shalen, and Johannson, whose work served to define and prove the existence and uniqueness of JSJ decompositions. See [**71**] and [**72**].

Definition 3.8.1. A 3-manifold is *atoroidal* if it contains no essential torus submanifold.

Theorem 3.8.2. *A closed orientable irreducible 3-manifold M contains a collection of disjointly embedded incompressible tori $\mathcal{T}$ such that each component of the 3-manifold obtained by cutting M along $\mathcal{T}$ is either atoroidal or Seifert fibered. Moreover, a minimal such collection of tori is unique up to isotopy.*

A similar theorem holds for non-orientable 3-manifolds so long as one also includes Klein bottles in the collection $\mathcal{T}$.

Definition 3.8.3. The collection of tori $\mathcal{T}$ as above is called the JSJ decomposition of M.

The diagram in Figure 3.39 describes a large class of 3-manifolds with similar JSJ decompositions. Each edge corresponds to a torus, each vertex to either a Seifert fibered space or an atoroidal 3-manifold. For instance, vertex 1 might correspond to an atoroidal 3-manifold with two torus boundary components. Vertices 2 and 3 might correspond to Seifert fibered spaces with at least three torus boundary components. A pair of the boundary components of each of these two Seifert fibered spaces is identified; another boundary component is identified to a boundary component of the atoroidal 3-manifold corresponding to vertex 1.

3.8. JSJ Decompositions   97

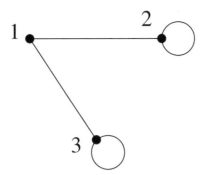

Figure 3.39. *Schematic for a JSJ decomposition.*

Definition 3.8.4. If the result of cutting M along $\mathcal{T}$ yields only Seifert fibered spaces, then M is called a *graph manifold*.

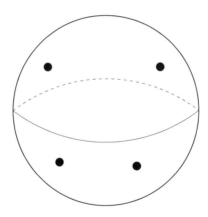

Figure 3.40. *A JSJ decomposition of a graph manifold.*

Figure 3.40 can be interpreted as a graph manifold. For instance, imagine two Seifert fibered spaces, M_1, M_2, each with base space a disk and two exceptional fibers glued along their boundaries. The great circle indicated corresponds to a saturated torus. If the gluing of M_1 and M_2 is performed in such a way that fibers do not match up with fibers, even up to isotopy, then the great circle indicated corresponds to the unique torus in the JSJ decomposition. If the gluing is such that fibers do match up with fibers, then the result is a Seifert fibered space with base space the sphere and four exceptional fibers; hence the JSJ decomposition is empty.

Theorem 3.8.2 can be rephrased as follows: "A closed orientable irreducible 3-manifold M has a Seifert fibered submanifold with atoroidal complement. The Seifert fibered submanifold with the least possible number of boundary tori is unique up to isotopy."

Definition 3.8.5. The Seifert fibered submanifold of a closed 3-manifold M that has the least possible number of boundary tori is called the *characteristic submanifold* of M.

Consider the graph manifolds described above. In the first case, the characteristic submanifold has two components: (shrunk versions of) M_1 and M_2. In the second case, the characteristic submanifold is the entire manifold.

Exercises

Exercise 1. Construct a graph manifold with three tori in its JSJ decomposition.

Exercise 2. By considering the 3-torus, show that the minimality assumption in Theorem 3.8.2 is necessary.

3.9. Compendium of Standard Arguments

There are several constructive arguments that have become standard in low-dimensional topology. We have used some of these in this chapter.

Example 3.9.1. *Standard innermost disk argument*: Building on Lemma 3.2.3, one can isotope incompressible surfaces in an irreducible 3-manifold so that their intersection contains only simple closed curves that are essential in the two incompressible surfaces. Indeed, suppose there are simple closed curves in the intersection of two incompressible surfaces, S_1 and S_2, in the 3-manifold M, that are inessential in, say, S_1. Let c be an inessential simple closed curve in $S_1 \cap S_2$ that is innermost in S_1. Then at least one component of $S_1 \backslash c$ is a disk D that is disjoint from S_2. Since c bounds a disk in M, it also bounds a disk D' in S_2. Since M is irreducible, the 2-sphere $D \cup D' \cup c$ bounds a 3-ball in M. Thus by Lemma 3.2.3 there is an isotopy reducing the number of components of $S_1 \cap S_2$. See Figure 3.41.

In particular, if one of the incompressible surfaces is a sphere or disk, then a succession of such isotopies eliminates all closed curves of intersection. We used this argument in the proofs of Theorem 3.2.5, Theorem 3.2.7, Lemma 3.7.14, and Theorem 3.7.17.

Lemma 3.9.2. *Let S_1, S_2 be surfaces in a 3-manifold M. Suppose that $S_1 \cap S_2$ contains a simple arc a such that one component of $S_i \backslash a$ is a disk D_i such that D_2 is disjoint from S_1. If $D_1 \cup D_2$ bounds a 3-ball in M, then:*

- *$(S_1 \backslash D_1) \cup D_2$ is isotopic to S_1;*
- *there is an isotopy of S_1 that eliminates the curve c from $S_1 \cap S_2$ and introduces no new components of intersection.*

3.9. Compendium of Standard Arguments 99

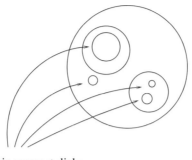

innermost disks

Figure 3.41. *Innermost disks.*

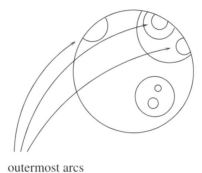

outermost arcs

Figure 3.42. *Outermost arcs.*

The proof of this lemma is analogous to the proof of Lemma 3.2.3.

Example 3.9.3. *Standard outermost arc argument*: Building on Lemma 3.9.2, one can isotope a pair of incompressible and boundary incompressible surfaces in an irreducible 3-manifold so that their intersection contains no inessential arcs. We did not use the standard outermost arc argument in this pure form.

Occasionally, this argument can be applied under weaker hypotheses to eliminate arcs of a specified type or to describe isotopies. We did not use this weaker form of the standard outermost arc argument in this chapter. Echos of the standard outermost arc argument can be seen for instance toward the end of the proof of Lemma 3.7.14.

Example 3.9.4. *Cut and paste arguments*: These are not as standard as the other two types of arguments. Recall the usage of the word cut in Definition 3.2.4. Likewise, *paste* refers to taking a union of two manifolds along their boundary or frontier. Specifically, we create an identification space from two spaces X, Y by identifying pairs of points $\mathbf{x} = [x, y]$, where x, y lie in specified subsets of X, Y. We also say that we *glue* X and Y along the specified subsets, especially if there is a function $f : X \to Y$ such that $y = f(x)$, when we say that we *glue* X to Y via f.

The ideas of cutting and pasting are implicit in constructions such as, e.g., $(S_1 \backslash D_1) \cup D_2$ in Lemmas 3.2.3 and 3.9.2. Standard innermost disk and outermost arc arguments are collectively referred to as cut and paste arguments. Cut and paste arguments can involve the Euler characteristic. Here it is important to note that, for instance, the replacement of S_1 by $(S_1 \backslash D_1) \cup D_2$ in Lemmas 3.2.3 and 3.9.2 has no impact on the Euler characteristic of the surface under consideration.

Chapter 4

Knots and Links in 3-Manifolds

Knots provide an excellent starting point for the understanding and appreciation of 3-manifolds. My generation of 3-manifold theorists found challenge and joy in working through Rolfsen's inspired and inspiring book [**127**]. Many of the standard constructions used in the topological study of knots and presented in Rolfsen's book rely on underlying general principles pertaining to 3-manifolds. They also represent key examples illustrating these principles. For this reason we include a brief introduction to the beautiful world of knots in 3-manifolds.

For a gentle introduction to knot theory, see [**1**] or [**89**]. For a more rigorous treatment, see [**19**] or [**87**].

4.1. Knots and Links

In general, a knot is an equivalence class of submanifolds. Our definitions (of submanifold, simple closed curve, knot, and link) rule out what are called "wild knots". For example, see Figure 4.1.

Figure 4.1. *A wild knot.*

Definition 4.1.1. A *knot* in $\mathbb{S}^3$ is a smooth isotopy class of smooth embeddings of $\mathbb{S}^1$ into $\mathbb{S}^3$. We write K to denote such an equivalence class.

We think of a representative of a knot as a simple closed curve k in $\mathbb{S}^3$. Two simple closed curves k, k' represent the same knot, i.e., are *equivalent*, if there is an isotopy of pairs between $(\mathbb{S}^3, k)$ and $(\mathbb{S}^3, k')$. Note that in Definition 4.1.1 we are not considering orientations on $\mathbb{S}^1$ or k. In particular, k and $-k$ represent the same knot.

Definition 4.1.2. An *n-component link* in $\mathbb{S}^3$ is a smooth isotopy class of smooth embeddings of the union of n copies of $\mathbb{S}^1$ into $\mathbb{S}^3$. We write L to denote such an equivalence class.

See the examples in Figures 4.2, 4.3, 4.4, and 4.5.

More generally, we consider smooth isotopy classes of simple closed curves in a given 3-manifold. Indeed, we will do so in the applications in later chapters of this book. However, for the purposes of this chapter, we restrict our attention to knots and links in $\mathbb{S}^3$. Moreover, since $\mathbb{R}^3 = \mathbb{S}^3 \setminus \{\text{point}\}$, this is equivalent to considering knots in $\mathbb{R}^3$.

Figure 4.2. *The unknot.*

Definition 4.1.3. The *unknot* is the isotopy class that contains the unknotted circle $\mathbb{S}^1 = \{(x, y, z) \mid x^2 + y^2 = 1, z = 0\} \subset \mathbb{R}^3$.

Henceforward, we will blur the distinction between a knot and its representatives.

In depicting knots, we consider projections of knots in $\mathbb{R}^3$ onto a plane. An application of transversality guarantees that in this projection we need only worry about two points projecting to the same point (rather than three or more) and that there will be only a finite number of such double points. At such double points, our convention is for instance to draw the arc containing the points above and closer to the plane as a broken line and the arc containing the points above and further away from the plane as a solid arc. See Figures 4.3 and 4.4. In this fashion, we capture all information about the knot in the planar projection. In other words, the projection can

4.1. Knots and Links

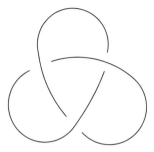

Figure 4.3. *The trefoil.*

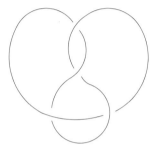

Figure 4.4. *The figure 8 knot.*

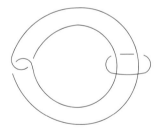

Figure 4.5. *The Whitehead link.*

be interpreted as a depiction of the knot in $\mathbb{S}^3$ in which all but a few short arcs lie in the plane. We call this type of projection a *knot diagram*.

Since a knot is an isotopy class and due to the fact that there are many possible projections of a knot in $\mathbb{S}^3$, diagrams of knots that look different might very well depict the same knot. Indeed, to create very complicated diagrams of the unknot, proceed as follows: Take a pencil and start drawing a curve. When you return to the curve, pass under the curve drawn so far to create an undercrossing in the projection. Continue in this fashion, passing under the (broken) curve each time you return. You can pass under the (broken) curve any number of times and wind your way around the page any number of times. Then return to the starting point to close up and create a knot projection. See Figure 4.6.

Figure 4.6. *A complicated diagram of the unknot.*

To see that the above description yields the unknot, imagine that your pen is moving farther and farther back as you draw. The process is consistent with this interpretation. At the very end, you add a short arc from way far back to the point where you started that is perpendicular to the planar projection you are considering. Now imagine pulling the former part of the knot tight, so that it becomes parallel to the latter arc. Then you see that the closed curve drawn is the unknot.

The two main questions in knot theory are the following:

Question 4.1.4. How do you tell whether or not two knot diagrams represent the same knot?

Question 4.1.5. How do you generate a list of all possible knots?

All knots are homeomorphic. They are all circles. What is important is the precise way that a circle sits in $\mathbb{S}^3$. Note that if instead we were to consider smooth isotopy classes of smooth embeddings of $\mathbb{S}^1$ into $\mathbb{S}^2$, then our task would be rather easy and unrewarding. In this case the Jordan Curve Theorem tells us that for 1-dimensional knots in $\mathbb{S}^2$ there is only one knot, namely the unknot. Likewise, if we were to consider smooth isotopy classes of smooth embeddings of $\mathbb{S}^2$ into $\mathbb{S}^3$, then our task would also be rather easy and unrewarding. In this case the Schönflies Theorem tells us that for 2-dimensional knots in $\mathbb{S}^3$ there is only one possibility, namely the standard 2-sphere sitting in $\mathbb{S}^3$.

Interestingly enough, the same phenomenon occurs for different reasons if we consider smooth isotopy classes of smooth embeddings of $\mathbb{S}^1$ into $\mathbb{S}^4$. Consider for instance the diagram of the trefoil in Figure 4.3. This diagram represents a knot that sits in a 3-dimensional subspace (in fact, in a small 3-dimensional neighborhood of a 2-dimensional subspace) of $\mathbb{S}^4$. At a crossing, imagine pushing the subarc of the undercrossing off into the fourth

4.1. Knots and Links

dimension. Then in a parallel 3-dimensional subspace we isotope the arc into an overcrossing arc. Now we isotope it back into our original 3-dimensional subspace. This way we can switch crossings so that an overcrossing becomes an undercrossing and vice versa.

By changing overcrossings to undercrossings we can turn any given knot diagram into one obtained by the above prescription for drawing a diagram of the unknot. In other words, the diagram is changed into one representing the unknot. This means that any smooth embedding of $\mathbb{S}^1$ into $\mathbb{S}^4$ represents the unknot. Thus for 1-dimensional knots in $\mathbb{S}^4$ there is also only one knot, again the unknot.

Finally, we introduce an infinite but fairly well-understood class of knots. These knots live on the torus and for this reason are called torus knots.

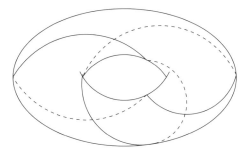

Figure 4.7. The $(3,2)$-torus knot.

Definition 4.1.6. Recall the definition of the n-torus in Example 1.1.5. In $\mathbb{R}^2$, consider the line $y = \frac{p}{q}x$ for $p, q \in \mathbb{Z}$ and $(p,q) = 1$ (i.e., p and q are relatively prime). The covering map $\pi : \mathbb{R}^2 \to \mathbb{T}^2$ maps this line to a simple closed curve on the torus. We call this simple closed curve a *torus knot*, more specifically, the (p,q)-*torus knot*. See Figure 4.7.

The 2-torus can be embedded in $\mathbb{S}^3$ in such a way that it bounds a solid torus on either side. This embedding is called the *unknotted solid torus*. Given a torus knot, it then determines a knot in $\mathbb{S}^3$ that is also called a *torus knot* and more specifically the (p,q)-*torus knot*.

In the context of 3-manifolds we are often interested in the knot complement $C(K) = \mathbb{S}^3 - \eta(K)$, where $\eta(K)$ is an open regular neighborhood of K. One of the most fundamental theorems in the subject of knot theory (and one that was rather difficult to prove) is the following (see [**46**]):

Theorem 4.1.7 (Gordon-Luecke). *Knots are determined by their complement. (I.e., if $C(K)$ and $C(K')$ are homeomorphic, then K and K' are isotopic.)*

The boundary of $C(K)$ is a single torus. It is the boundary of a closed regular neighborhood of the knot. A simple closed curve on $\partial C(K)$ that bounds a disk in $\eta(K)$ is called a *meridian*. Any simple closed curve that intersects the meridian once is called a *longitude*. All meridians are isotopic. All longitudes are not isotopic. In the case of the unknot the *preferred* longitude is the curve that bounds a disk in the complement $C(K)$ of the unknotted solid torus $\eta(K)$. More generally, the preferred longitude is the curve that bounds a compact connected orientable surface with a single boundary component. Such a surface always exists and is called a Seifert surface. We will define and discuss these in Section 4.3.

Note that the meridian and the preferred longitude for the unknot, (m, l), provide a coordinate system for torus knots: Specifically, given a representative k of the (p, q)-torus knot, $p = [m] \cdot [k]$ and $q = [l] \cdot [k]$.

Exercises

Exercise 1. Construct a knot out of a piece of string and experiment by moving it around in space, tugging first on one portion and then another.

Exercise 2. Draw several diagrams based on projections of the knot in Exercise 1.

Exercise 3. Contemplate criteria to distinguish the unknot from the trefoil.

Exercise 4. Exhibit the trefoil as a torus knot by finding appropriate p and q.

Exercise 5. Prove that the complement of a torus knot is a Seifert fibered space.

4.2. Reidemeister Moves

Manipulating knot diagrams can be a daunting task. It requires some amount of visual acuity. The work of Reidemeister and independently Alexander and Briggs in the 1920s streamlined this task significantly. The idea was to formalize the procedure of showing that two diagrams represent the same knot. This was achieved by isolating three moves. These are now called the *Reidemeister moves*. See Figures 4.8, 4.9, and 4.10.

These moves represent isotopies that take place in a specified region of the knot diagram. Inside the region the diagram looks as pictured on the left-hand side (or the right-hand side, respectively) of one of the Figures 4.8, 4.9, or 4.10. After the move, the inside of the region looks as pictured on the right-hand side (or the left-hand side, respectively) of the same figure. Everything outside of the specified region remains the same.

4.2. Reidemeister Moves

Figure 4.8. *Type I Reidemeister move.*

Figure 4.9. *Type II Reidemeister move.*

Figure 4.10. *Type III Reidemeister move.*

Definition 4.2.1. A *planar isotopy* of a knot diagram is an isotopy of the arcs in the knot diagram that does not affect the crossings.

What makes the Reidemeister moves so effective is the following theorem:

Theorem 4.2.2 (Reidemeister). *Two knot diagrams represent the same knot if and only if one diagram can be transformed into the other by a sequence of Reidemeister moves and planar isotopies.*

In many areas of knot theory the Reidemeister moves provide the tool to prove that a property of knots or a quantity is well-defined. This is especially true for many of the modern knot invariants. We will not pursue this here but will provide one illustration of how to use the Reidemeister moves in this way.

Definition 4.2.3. A knot diagram is *3-colorable* if the arcs in the diagram can be colored in such a way that the following hold: (1) Each arc in the diagram is colored with a single color; (2) at each crossing, either all arcs meeting at that crossing are colored with the same color or all three colors are used; (3) all three colors are used in the diagram.

Theorem 4.2.4. *The property of being 3-colorable is well-defined for knots.*

Proof. We need to show that if one diagram of a knot is 3-colorable, then so is any other diagram. So suppose a diagram of the knot is 3-colorable.

Any other diagram of the knot is obtained from this diagram by a sequence of Reidemeister moves and planar isotopies. Now observe that if a diagram is 3-colorable, then it is also 3-colorable after a Type I Reidemeister move because only one color can occur on a strand and also at the crossing of the loop. The same is true after a Type II Reidemeister move. Indeed, either all strands are the same color before and after the Reidemeister move or all three colors occur in the strands on the left-hand side and only two colors occur on the right-hand side of Figure 4.9. If this is the case, then since two colors occur on the right-hand side and since the knot is connected, there must be a crossing involving all three colors. The situation for a Type III move is even simpler. □

The property of being 3-colorable provides a means of distinguishing between knots based on their diagrams. If we compare two knot diagrams and find that one is 3-colorable whereas the other is not, then we know that the two diagrams represent distinct knots.

Exercises

Exercise 1. Find a sequence of Reidemeister moves that transforms the diagram of the unknot pictured in Figure 4.6 into the round unknot.

Exercise 2. Prove that the unknot and the trefoil are distinct using Theorem 4.2.4.

4.3. Basic Constructions

Lay people confronted with knot theory often ask what knots could possibly have to do with mathematics, since they equate mathematics with algebra. Their peace of mind can be restored by an appeal to the semigroup properties enjoyed by knots. Indeed, there is an addition operation for knots. In the definition of this addition operation we will think of a knot as an oriented 1-dimensional submanifold of the oriented 3-manifold $\mathbb{S}^3$, with orientation inherited from the embedding of $\mathbb{S}^1$ with the standard orientation.

Definition 4.3.1. Let $(\mathbb{S}^3, K_1)$, $(\mathbb{S}^3, K_2)$ be oriented knots. Their *connected sum* is the oriented pairwise connected sum $(\mathbb{S}^3, K_1) \# (\mathbb{S}^3, K_2)$.

Remark 4.3.2. Abusing notation slightly, we refer to our knots as K_1 and K_2 and simply write $K_1 \# K_2$.

Figure 4.11 depicts the connected sum of the trefoil (see Figure 4.3) and the figure 8 knot (see Figure 4.4). In the oriented pairwise connected sum we remove from each pair a 3-ball of $\mathbb{S}^3$ containing a subarc of K_i. We then identify the resulting 2-sphere boundary components in such a way that the

4.3. Basic Constructions

Figure 4.11. *The connected sum of the figure 8 knot and the trefoil.*

endpoints of the deleted arcs are identified. A priori there are two choices, up to isotopy, for this identification, but we decree that the orientations match up. In particular, the endpoints of the arcs must be identified in such a way that the induced orientations on the subarcs match up.

In the unoriented category this operation is not well-defined. See Figures 4.12 and 4.13. Moreover, this addition operation is not well-defined for links.

Figure 4.12. *The square knot.*

Figure 4.13. *The granny knot.*

The unknot acts as the identity in the semigroup defined by this addition operation. Note that the operation is commutative. There are no inverses for this operation. This is a non-trivial fact that we leave as an exercise. The definitions below and Theorem 4.3.6 provide tools that suffice to establish this fact.

Definition 4.3.3. A *Seifert surface* of a knot K is a compact connected orientable surface S in $\mathbb{S}^3$ such that $\partial S = K$.

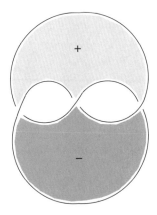

Figure 4.14. *A Seifert surface for the trefoil knot.*

Theorem 4.3.4. *Every knot possesses a Seifert surface.*

Proof (Seifert's Algorithm).

Step 1. Give the knot an orientation.

Step 2. Examine a diagram of the knot. At each crossing, alter the diagram as in Figure 4.15.

Figure 4.15. *Replacement at each crossing.*

The result is a collection of disjoint oriented simple closed curves.

Step 3. Consider disks in $\mathbb{R}^2$ bounded by the simple closed curves. Orient these disks with a + or a − depending on whether their boundaries are oriented counterclockwise or clockwise, respectively. If curves are nested, then the disks can be made disjoint by raising them out of the plane of the diagram with the innermost one raised to be highest.

Step 4. Connect the disks at the old crossings via half-twisted strips as in Figure 4.16.

4.3. Basic Constructions

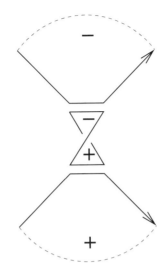

Figure 4.16. *Twisted band for each crossing.*

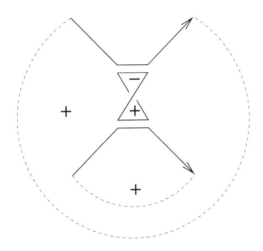

Figure 4.17. *Twisted band for each crossing.*

If a twisted band connects non-nested disks, then one disk is oriented with a + and the other with a −. So attaching the twisted band is consistent with the orientations of the disks. See Figure 4.16.

If a twisted band connects nested disks, then either both disks are oriented with a + or both are oriented with a −. Here too, attaching the twisted band is consistent with the orientations of the disks. See Figure 4.17. (Note that the band lies above the lower disk and the − side of the band connects to the underside of the lower disk.) □

Seifert surfaces capture some of the quality of the knottedness exhibited by a knot. They also provide an important tool in many investigations of knot complements. In some cases, they help us distinguish between knots by means of the following definition:

Definition 4.3.5. The *genus* of a knot K, denoted by $g(K)$, is the smallest possible genus of a Seifert surface of K.

Theorem 4.3.6. *Let K_1, K_2 be knots. Then $g(K_1 \# K_2) = g(K_1) + g(K_2)$.*

Proof. To prove this equality, we prove two inequalities. First we prove the following inequality:
$$g(K_1 \# K_2) \leq g(K_1) + g(K_2).$$

Consider minimal genus Seifert surfaces S_1 and S_2 for K_1 and K_2, respectively, and take the connected sum of triples for $(K_1, S_1, \mathbb{S}^3)$ and $(K_2, S_2, \mathbb{S}^3)$. In doing so we remove from S_i a small disk with boundary partitioned into an arc in ∂S_i and an arc in the interior of S_i and glue the surfaces along the two interior arcs. As the result, we obtain a Seifert surface of $K_1 \# K_2$ of genus $g(K_1) + g(K_2)$. This gives an upper bound for the genus of $K_1 \# K_2$.

Now we prove the other inequality:
$$g(K_1 \# K_2) \geq g(K_1) + g(K_2).$$

Consider a minimal genus Seifert surface S for $K_1 \# K_2$. We wish to use S to construct Seifert surfaces for K_1 and K_2. As we will see, this is achieved by the cut and paste techniques typical of 3-manifold topology.

It follows from the definition of connected sum of knots that there is a 2-sphere Z in $\mathbb{S}^3$ that separates $K_1 \# K_2$ into its two summands. In particular, the knot intersects Z in exactly two points. The intersection of S with Z, after an isotopy making S and Z transverse, consists of simple closed curves and simple arcs. Moreover, since the endpoints of arcs correspond to points of intersection with the knot, there must be exactly one arc of intersection. The complement of this arc of intersection in Z is a disk D containing only closed components of intersection.

Suppose first that there are no closed components of intersection. Then the arc of intersection cuts S into two components, one, S_1, a Seifert surface for K_1, the other, S_2, a Seifert surface for K_2. Furthermore, $g(S_1) + g(S_2) = g(S)$, so the above inequality holds.

You will prove in the exercises that the complement of a knot in $\mathbb{S}^3$ is irreducible and that a minimal genus Seifert surface is incompressible. Thus closed components of $S \cap Z$ can be removed via a standard innermost disk argument. Now the above construction yields the desired inequality. □

Exercises

Exercise 1*. Prove that K is the unknot if and only if $g(K) = 0$. (It is not hard to see that if K is the unknot, then $g(K) = 0$. The converse is challenging. Hints: If $g(K) = 0$, consider a regular neighborhood $N(K)$ and the Seifert surface of K that is a disk D. Use these to describe $\mathbb{S}^3$ as a lens space. Use the fact that $\mathbb{S}^3$ has trivial fundamental group to show that K must be the unknot.)

Exercise 2. Prove that if K_1 and K_2 are non-trivial knots, then $K_1 \# K_2$ can never be the unknot. Deduce that there are no inverses for the operation of connected sum. (Hint: Use Exercise 1 and Theorem 4.3.6.)

Exercise 3. Prove that every knot has a prime decomposition. (This prime decomposition is unique up to reordering of summands, but this is harder to prove.)

Exercise 4. Construct Seifert surfaces that show that the genus of the (p, q)-torus knot is at most $\frac{1}{2}(p-1)(q-1)$. (In fact, the genus of the (p, q)-torus knot is $\frac{1}{2}(p-1)(q-1)$, though this is harder to prove.)

Exercise 5. Prove that a minimal genus Seifert surface of a knot in $\mathbb{S}^3$ is incompressible.

Exercise 6. Prove that the complement of a knot in $\mathbb{S}^3$ is irreducible.

4.4. Knot Invariants

The genus of a knot is an example of a knot invariant. More generally, a *knot invariant* is any quantity that is well-defined on the isotopy class of the knot. The property of 3-colorability is another example of a knot invariant. In recent decades, the study of knot invariants has taken on a very different flavor than the study of 3-manifolds. Many beautiful algebraic insights have been obtained, for instance, in the study of topological quantum field theory and more recently in the study of Heegaard Floer homology. Discussing these results would take us too far afield from our goals here. Suffice it to say that recent years have witnessed a resurgence by the more algebraically inclined to understand the topological implications of these insights and to interpret such insights for a more topologically inclined audience.

We wish to discuss some of the more classical knot invariants. We have already seen the property of 3-colorability and the genus of a knot. The following definition gives us a very natural invariant that relies heavily on the representation of knots via diagrams:

Definition 4.4.1. The *crossing number* of a knot K, denoted by $c(K)$, is the least number of crossings in a diagram of the knot.

Example 4.4.2. The crossing number of the trefoil is 3.

To prove that the crossing number of the trefoil is indeed 3, we must prove two things: (1) that the crossing number of the trefoil is at most 3; (2) that the crossing number of the trefoil is no smaller than 3.

To prove the first statement, we simply observe that the diagrams of the trefoil exhibited in Figures 4.3 and 4.14 both have crossing number 3. So the crossing number of the trefoil is at most 3.

Proving the second statement requires more work. Why can't the crossing number of the trefoil be 0, 1, or 2? Fortunately, we can rule out these options one by one, though each requires a small amount of work. The crossing number of the trefoil cannot be 0 because a diagram with crossing number 0 necessarily depicts the unknot (this is the Jordan Curve Theorem). To show that it can't be 1, draw one crossing and try to connect up the endpoints of the arcs without introducing any other crossings. In all cases we obtain an unknot that can be transformed into the round unknot after one Type I Reidemeister move and planar isotopies. An analogous analysis shows that the crossing number of a non-trivial knot cannot be 2.

More generally, Murasugi proved the following (see [**110**]):

Theorem 4.4.3 (Murasugi). *The crossing number of the (p,q)-torus knot is*

$$\min\{p(q-1), q(p-1)\}.$$

Knot tables employ the crossing number as an ordering scheme. Knots with crossing number 3 are listed first, then knots with crossing number 4, then with crossing number 5, etc. Incidentally, there is only one knot, the trefoil, with crossing number 3 and only one knot, the figure 8 knot (see Figure 4.4), with crossing number 4. We leave the proofs of these facts as exercises.

The number of knots with a given crossing number is of interest. The sequence starts with the number of knots of crossing number 3 and continues on up. The first few terms are 1, 1, 2, 3, 7, 21, 49, 165, 552, 2176, 9988, 46972, ... This explosive growth shows that for some purposes, the crossing number is not a good ordering scheme.

Some interesting facts are known about the crossing number. We give one example below. But many natural questions remain.

4.4. Knot Invariants

Definition 4.4.4. A knot diagram is *alternating* if the over or under nature of the crossings alternates as one travels along the knot. A knot is *alternating* if it has an alternating diagram.

Definition 4.4.5. A *nugatory* crossing in a knot diagram is a crossing such that there is a circle that intersects this crossing and is disjoint from the rest of the knot diagram. See Figure 4.18.

Figure 4.18. *A nugatory crossing.*

Note that a nugatory crossing can be removed via an isotopy (in $\mathbb{S}^3$ and hence also by a series of Reidemeister moves).

Definition 4.4.6. A knot diagram is *reduced* if it contains no nugatory crossings.

A major result concerning crossing number was obtained independently by Kauffman, Murasugi, and Thistlethwaite:

Theorem 4.4.7 (Kauffman, Murasugi, and Thistlethwaite)**.** *A reduced alternating diagram of K realizes the minimal crossing number of K.*

Many questions concerning the crossing number remain. For instance, it is unknown how the crossing number behaves under the operation of connected sum. It is not too hard to show that the crossing number of a composite knot is at most the sum of the crossing numbers of its summands. We leave the proof of this fact as an exercise. Consider the implications of this fact in light of the number of knots of a given crossing number. The number of knots with a given crossing number grows explosively. One upshot is the interesting fact that the number of prime knots with a given crossing number is far greater than the number of composite knots with that crossing number. On the other hand, from the probabilistic point of view pursued, for instance, in [**36**], it follows, by [**35**], that a generic knot is composite, more specifically, has a trefoil summand.

Conjecture 4.4.8. *Crossing number is additive under connected sum of knots.*

Another classical knot invariant grows out of the following: Consider a diagram of K and change one of the crossings either from an overcrossing to an undercrossing or vice versa. This yields a knot K_1, typically distinct from K. Now consider a diagram of K_1 (not necessarily related to the diagram of K) and continue the process. For every knot, this process can be chosen in such a way that the process terminates with the unknot. To see this, recall our construction of complicated diagrams for the unknot. Using this idea, any knot diagram can be transformed into a diagram of the unknot by a finite number of crossing changes. Specifically, if the diagram has n crossings, we need to change at most $\frac{n}{2}$ crossings.

Definition 4.4.9. The *unknotting number* of a knot is the minimal number n such that K_n (in the above process) is the unknot. (The unknotting number is also called the *Gordian* number.) We denote the unknotting number of a knot K by $u(K)$.

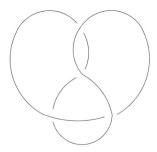

Figure 4.19. *A crossing change on a diagram of the figure 8 knot yielding the unknot.*

The discussion preceding the definition ensures that the unknotting number is well-defined. It also proves the following theorem:

Theorem 4.4.10. *For any knot K, $u(K) \leq \frac{1}{2}c(K)$.*

The unknotting number appears to be related to 4-dimensional phenomena. Consider the process of unknotting a knot K in $\mathbb{S}^3$. Over time, the knot sweeps out a surface in $\mathbb{S}^3 \times I \subset \mathbb{B}^4$. The questions and techniques studied in 4-dimensional topology differ greatly from those in 3-manifold topology. For instance, gauge theory has been an effective tool in the study of 4-manifolds. Using gauge theory, Kronheimer and Mrowka proved the following theorem (see [**82**]):

Theorem 4.4.11 (Kronheimer-Mrowka). *The (p,q)-torus knot has unknotting number $\frac{1}{2}(p-1)(q-1)$.*

4.4. Knot Invariants

Proving that the unknotting number of the (p,q)-torus knot is at most $\frac{1}{2}(p-1)(q-1)$ is comparatively easy. One need only consider an appropriate diagram of the (p,q)-torus knot. Kronheimer and Mrowka's contribution lay in proving the reverse inequality.

Despite its intuitive appeal, understanding the unknotting number remains tantalizingly out of reach. The big open question parallels the above question on crossing number.

Question 4.4.12. How does the unknotting number behave under the connected sum of knots?

Some amount of progress on this question was made by Scharlemann, who proved the following theorem:

Theorem 4.4.13 (Scharlemann). *Unknotting number* 1 *knots are prime.*

Yet another invariant for links is the linking number. To compute it, we must first orient the components of our link. Once each component is oriented, we can distinguish between positive and negative crossings. See Figure 4.20.

Figure 4.20. *A positive and a negative crossing.*

Now given a link, we can consider one-half of the difference of the number of positive and negative crossings.

Definition 4.4.14. The *linking number* between two components L_1, L_2 of a link is one-half of the number of positive crossings minus the number of negative crossings. It is denoted by $lk(L_1, L_2)$.

The linking number of the two components of the Hopf link is ± 1 depending on the orientations chosen. See Figure 4.21.

Recent years have seen rapid progress in understanding invariants of knots and 3-manifolds growing out of homology theories. These subjects go beyond the scope of this book. The reader is invited to consult [76], [116], [117], and [118].

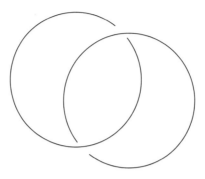

Figure 4.21. *The Hopf link.*

Exercises

Exercise 1. Show that the only knot with crossing number 3 is the trefoil.

Exercise 2. Show that the only knot with crossing number 4 is the figure 8 knot.

Exercise 3. Show that the crossing number of a composite knot is at most the sum of the crossing numbers of its summands.

Exercise 4. Show that the crossing number of the (p,q)-torus knot is at most $p(q-1)$.

Exercise 5. Show that the crossing number of an unknotting number 1 knot can be arbitrarily large.

Exercise 6. Show that the unknotting number of the (p,q)-torus knot is at most $\frac{1}{2}(p-1)(q-1)$.

Exercise 7. Show that the unknotting number of a composite knot is at most the sum of the unknotting numbers of its summands.

Exercise 8. Prove Conjecture 4.4.8 for alternating knots. (Hint: A literature search is appropriate.)

4.5. Zoology

We mentioned above that the crossing number can be an ineffective tool for ordering knots due to the explosive growth of the number of knots with a given crossing number. Instead, we often focus on certain classes or types of knots. We have already discussed one such class of knots, namely the torus knots. This is an infinite family of knots. Not only is this class of knots infinite, but it also includes knots of arbitrarily high crossing number and

4.5. Zoology

of arbitrarily high unknotting number. In this section we introduce a few more rich classes of knots.

A *pretzel knot* is a knot modeled on Figure 4.22. Here the boxes containing the letters p, q, r represent that number of half-twists of the appropriate strands of the knot. In Figure 4.23 we illustrate, as an example, the $(-2, 3, 7)$-pretzel knot.

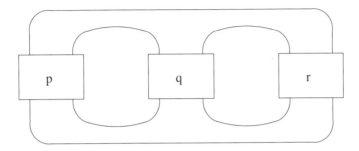

Figure 4.22. *Model for the (p, q, r)-pretzel knot.*

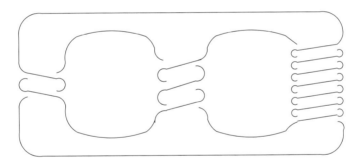

Figure 4.23. *The $(-2, 3, 7)$-pretzel knot.*

Pretzel knots are a subclass of a larger class called *Montesinos knots*. Montesinos knots, in turn, are a subclass of the class of *arborescent knots*. Arborescent knots have been classified by Bonahon and Siebenmann. The class of arborescent knots was also studied in detail by Gabai. (See [**42**].) Among other things, he found that for knots in this class the genus is the genus realized by a Seifert surface obtained from a "standard" diagram; in the case of a pretzel knot this would be Figure 4.23, via Seifert's algorithm.

A *2-bridge* knot is a knot that can be isotoped to have exactly two maxima with respect to a height function on $\mathbb{S}^3$. As it turns out, every 2-bridge knot is modeled on Figure 4.24. A box labeled t_i represents t_i half-twists of the appropriate strands of the knot. The trefoil and the figure 8 knot are examples of 2-bridge knots. 2-bridge knots also belong to the larger

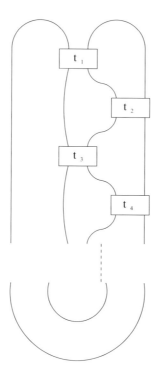

Figure 4.24. *Model for 2-bridge knots.*

class of arborescent knots. The repeated fraction expansion of a 2-bridge knot diagram as in Figure 4.24 is the rational number

$$\frac{p}{q} = \frac{1}{t_1 + \frac{1}{t_2 + \cdots \frac{1}{t_{n-1} + \frac{1}{t_n}}}}.$$

The repeated fraction expansion of a rational number $\frac{p}{q}$ is not unique, but as it turns out, 2-bridge knot diagrams correspond to the same knot if and only if they give repeated fraction expansions of the same rational number $\frac{p}{q}$. (In Conway's notation, the 2-bridge knot is simply the $\frac{p}{q}$-rational knot. See [28].) We will occasionally refer to 2-bridge knots in the chapters that follow. In certain settings 2-bridge knots and 2-bridge knot complements provide interesting examples of pathological behavior. In Section 4.8 below, we will discuss bridge numbers of knots, another knot invariant.

Given two knots, the operation of connected sum produces a new knot from the two given knots. In fact, the operation of connected sum is a special

4.5. Zoology

case of a more general binary operation:

Definition 4.5.1 (Satellite construction). Let A be a non-trivial knot in $\mathbb{S}^3$ and let V be a closed regular neighborhood of A. Furthermore, let B be a knot in the solid torus $\hat{V}$. The pair $(\hat{V}, B)$ is called the *pattern*. The minimal number of times that B intersects a meridian disk of $\hat{V}$ is called the *index* or *wrapping number*. Suppose that the index of B is at least 1. We denote the image of B under a homeomorphism of $\hat{V}$ into V by K. Here K is called a *satellite knot*. The knot A is called a *companion* of K. The boundary of V is called a *companion torus*.

Figure 4.25. *A satellite knot and its companion torus.*

The satellite construction is interesting for many reasons. One such reason is that the boundary of V is an essential torus in the complement of the satellite knot.

To reconstruct the connected sum of K_1 and K_2 from the satellite construction, set $A = K_1$ and choose a solid torus containing K_2 with wrapping number 1 as the pattern. Then the resulting satellite knot will be $K_1 \# K_2$. In this case the resulting essential torus is called a *swallow-follow torus*. (It "swallows" K_2 and "follows" K_1.) The roles of K_1 and K_2 can be reversed. This yields a different swallow-follow torus. (This one "swallows" K_1 and "follows" K_2.)

Remark 4.5.2. A knot can have many companion knots and companion tori. In the 1950s, Horst Schubert investigated the finiteness and relative positioning of companion tori; see [**141**]. In essence, he proved the existence and uniqueness of JSJ decompositions for knot complements.

Exercises

Exercise 1. Show that the plane in Figure 4.26 separates the figure 8 knot into trivial arcs. Specifically, draw a pair of disks above the plane and a pair of disks below the plane such that the pair exhibits an isotopy of the two arcs into the plane.

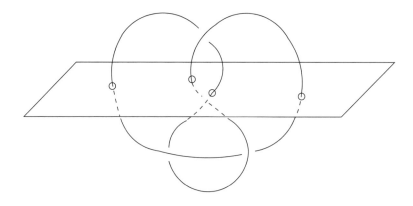

Figure 4.26. *A plane separating the figure 8 knot into trivial arcs.*

Exercise 2. Draw a swallow-follow torus for the connected sum of the trefoil and the figure 8 knot.

Exercise 3. Suppose that two non-isotopic swallow-follow tori have been isotoped to intersect in a minimal number of curves. What can you say about their intersection?

4.6. Braids

We now briefly discuss braids. For the purposes here, braids will merely be a special positioning for knots and links. But in fact, braids represent a point of view. Many fascinating results about knots and links have been obtained by using the structure given by the representation of knots and links as braids. Moreover, braids naturally give rise to a group. For this reason, braids have been studied not only by topologists, but also by algebraists and even mathematical physicists. For more on braids and related topics, see [**10**].

Definition 4.6.1. An *n*-string *braid* is a collection of disjoint arcs in a box that monotonically connect the top to the bottom. See Figure 4.27. Two braids are considered equivalent if they are isotopic in the box relative to their endpoints (i.e., via an isotopy that fixes the endpoints).

4.6. Braids

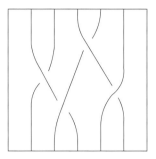

Figure 4.27. *A braid.*

Definition 4.6.2. Denote by σ_i the braid with a single positive crossing such that the i-th strand crosses over the $(i+1)$-th strand.

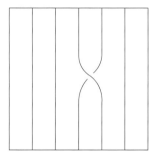

Figure 4.28. *The braid σ_3.*

Two n-string braids can be multiplied by stacking the first on top of the other. Under this operation, braids form a group called the *braid group*. This group has the presentation

$$\langle \sigma_1, \ldots, \sigma_{n-1} \mid \sigma_i \sigma_j = \sigma_j \sigma_i \text{ if } |i-j| \geq 2, \ \sigma_i \sigma_{i+1} \sigma_i = \sigma_{i+1} \sigma_i \sigma_{i+1} \rangle.$$

Definition 4.6.3. The *closure of a braid* is the diagram of a knot (or link) obtained from a braid by connecting the top endpoints of the arcs to the bottom endpoints of the arcs with disjoint arcs (without crossings). See Figure 4.29. Any knot (or link) in this form is called a *closed braid*.

Definition 4.6.4 (Alternate definition of closed braid)**.** An *axis* in $\mathbb{S}^3$ is an unknot A in $\mathbb{S}^3$ together with a product structure (open disk) $\times\, \mathbb{S}^1$ on $\mathbb{S}^3 \setminus A$. A *closed braid* is a link L in $\mathbb{S}^3$ for which there is an axis such that L intersects each (open disk) $\times$ {point} in the product structure transversely.

We leave it as an exercise to show how the two definitions of closed braid are equivalent. Observe that the depiction of the trefoil in Figure 4.3 is a closed braid whereas the depiction of the trefoil in Figure 4.30 is not.

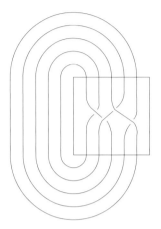

Figure 4.29. *A closed braid.*

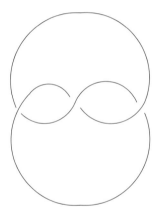

Figure 4.30. *The trefoil.*

Theorem 4.6.5 (Alexander)**.** *Every knot (or link) can be isotoped to be a closed braid.*

Sketch of proof; for details see [7]. Let L be an oriented link in $\mathbb{S}^3$. We will work in the TRIANG category in the sense that we will think of L as a finite collection of edges joined end to end. Consider a diagram of L in the plane.

Choose a point a in $\mathbb{R}^2$ (representing an axis in $\mathbb{S}^3$ orthogonal to $\mathbb{R}^2$) that is not collinear with any edge of L. Connect the endpoints of the edges of L with a to form triangles. See Figure 4.32. The orientation of L provides an orientation of the edges of L. Thus each such triangle inherits an orientation. We leave it as an exercise to show that L is a braid (in the

4.6. Braids

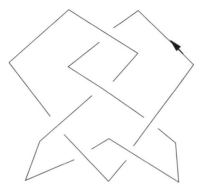

Figure 4.31. *The 6_1 knot.*

sense of Definition 4.6.4) if and only if all triangles coming from the same component of L have the same orientation.

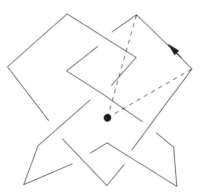

Figure 4.32. *A triangle.*

It now suffices to show that we can perform isotopies of L after which an arbitrary such triangle has the appropriate orientation. If a triangle has the wrong orientation and the edge $[AB]$ of L in the triangle contains no crossing points, choose a point $C \in \mathbb{R}^2$ as in Figure 4.33 such that the triangle $[ABC]$ contains a. Then replace the edge $[AB]$ with the two edges $[AC], [BC]$. This process may introduce new crossings of L, but as long as $[AC], [BC]$ always cross over (or always cross under) any other portions of L, the exchange can be accomplished via an isotopy of L. See Figure 4.33.

If the edge contains one overcrossing (undercrossing, respectively), an identical replacement can be made so long as at the new edges $[AC], [BC]$ always cross over (under, respectively) any other portions of L. If the edge contains multiple crossings, subdivide it into shorter edges, each containing exactly one crossing, and proceed analogously. □

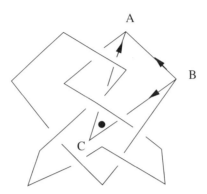

Figure 4.33. *A replacement.*

Exercises

Exercise 1. Write down the group element for the braid in Figure 4.27.

Exercise 2. Write down the group element of the braid whose closure is the (p,q)-torus knot.

Exercise 3. Find a closed braid representation of the figure 8 knot.

Exercise 4. Show that the two definitions of closed braid are equivalent. (Hint: Given a closed braid satisfying the alternate definition, "straighten" the arcs on the left and "squeeze" the portion containing crossings into a box.)

4.7. The Alexander Polynomial

The Alexander polynomial is the oldest known knot polynomial. We discuss it briefly as an exercise in understanding the ambient space of a knot, a 3-manifold. The definition is a lengthy one. We treat it only informally. Given a knot, there are several ways to compute a polynomial according to specific instructions. These several ways correspond to several different knot polynomials. In the last two decades, the Jones polynomial has eclipsed the Alexander polynomial in importance. The Jones polynomial is computed from a knot diagram. The connection to the ambient 3-manifolds is obscured. For this reason we omit a discussion of the Jones polynomial here. For the interested reader, Lickorish's book (see [**87**]) provides a rich introduction to the topic. In recent years, a generalization of the Jones polynomial, called the colored Jones polynomial, has been used in the formulation of the "Kashaev Conjecture". This conjecture postulates that a certain growth rate of the coefficients of the colored Jones polynomial gives the hyperbolic volume of the knot complement.

4.7. The Alexander Polynomial

To compute the Alexander polynomial of a knot K, we first consider a Seifert surface S of K. Here S is a once-punctured compact connected orientable surface. As such it is homeomorphic to the surface pictured in Figure 4.34.

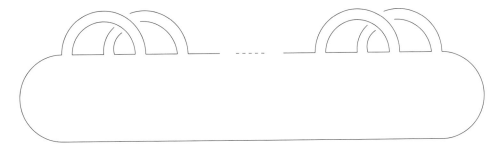

Figure 4.34. *The once-punctured compact orientable surface.*

We consider the collection of simple closed curves in this surface pictured in Figure 4.35. Those familiar with algebraic topology will recognize that the curves correspond to a set of generators for the first homology of the surface. The homeomorphism between S and the surface pictured in Figure 4.34 defines a collection of simple closed curves in S. (They will depend on the homeomorphism.)

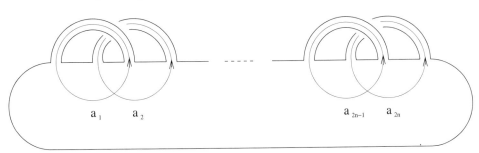

Figure 4.35. *A collection of oriented circles in the once-punctured compact orientable surface.*

Because S is an oriented submanifold of the oriented 3-manifold $\mathbb{S}^3$, Theorem 3.1.14 tells us that a regular neighborhood of S is a trivial bundle $S \times [-1, 1]$. Consider a copy of one of the circles a_i. We denote a copy of a_i that has been isotoped slightly off of S towards $S \times \{1\}$ by a_i^+. We call a_i a *push-off* of a_i. Figure 4.36 depicts the push-offs in the case of the trefoil.

Define a matrix A with ij-th entry equal to $(lk(a_i, a_j^+))$. The Alexander polynomial is the polynomial $\Delta_K = \det(tA - A^t)$, where A^t is the transpose of A. It is a non-trivial fact that this is well-defined up to sign, despite the numerous choices made.

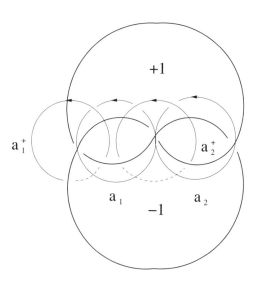

Figure 4.36. The circles a_1, a_2 and their push-offs a_1^+, a_2^+ for the trefoil.

Example 4.7.1. In the case of the trefoil, the computation is as follows:

$$A = \begin{bmatrix} -1 & 1 \\ 0 & -1 \end{bmatrix},$$

$$tA - A^t = \begin{bmatrix} -t & t \\ 0 & -t \end{bmatrix} - \begin{bmatrix} -1 & 0 \\ 1 & -1 \end{bmatrix} = \begin{bmatrix} -t+1 & t \\ -1 & -t+1 \end{bmatrix},$$

$$\Delta_K = t^2 - 2t + 1 + t = t^2 - t + 1.$$

If we had made other choices, we could have ended up with $-t^2 + t - 1$.

Exercises

Exercise 1. Convince yourself that every once-punctured compact connected orientable surface is homeomorphic to a surface as in Figure 4.34.

Exercise 2. Compute the Alexander polynomial for the figure 8 knot.

Exercise 3. Prove that if K_1 and K_2 are knots, then $\Delta_{K_1 \# K_2} = \Delta_{K_1} \Delta_{K_2}$.

4.8. Knots and Height Functions

Recall that a height function is a Morse function with at most two critical points. A height function h on $\mathbb{S}^3$ has a maximum and a minimum. We may assume that a knot K in $\mathbb{S}^3$ misses the maximum and the minimum. It then meets only regular level surfaces of h. Each regular level surface is a

4.8. Knots and Height Functions

2-sphere. Locally, we visualize K in $\mathbb{R}^3$ with h given by projection onto the z-coordinate. When we do so, we refer to this height function as the *natural height function*.

In the 1950s Schubert used height functions for some of his groundbreaking results on knots and knot complements. See [142] and [141]. In the process, he introduced several notions, for instance the notion of bridge number. The following is a reformulation of the notion of bridge number:

Definition 4.8.1. The *bridge number* of a knot K, denoted by $b(K)$, is the least number of maxima for K with respect to a height function on $\mathbb{S}^3$.

To motivate the terminology, consider the following alteration: Isotope all maxima of the knot K upward so that they all have the same height. Then consider a level surface L just below the maxima. Here L cuts K into two collections of arcs: those above L that each contain exactly one maximum and no other critical points and those below L that each contain exactly one minimum and no other critical points.

Consider the portion of K lying below L. It consists of subarcs that can be isotoped into L relative to their endpoints. The result of this isotopy is a collection of arcs Γ in L. Moreover, this isotopy can be performed in such a way that the resulting arcs Γ in L are disjoint. Now we consider a copy of K that consists of Γ along with the small arcs containing the maxima. Each of the latter arcs is a "bridge", hence the terminology. See Figure 4.37.

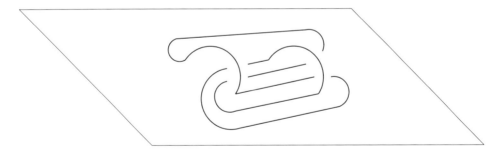

Figure 4.37. *Classical bridge position for the figure 8 knot.*

Definition 4.8.2. If K is embedded in $\mathbb{S}^3$ in such a way that all maxima occur above all minima of K, then K is said to be in *bridge position*.

In Section 4.5 we discussed 2-bridge knots. These knots derive their name from the fact that they are precisely the knots that have bridge number 2. (They include the trefoil and the figure 8 knot.) The embedding indicated in Figure 4.24 is in bridge position with respect to the natural height function. Note how the crossing number of a 2-bridge knot can be arbitrarily large. To see this, recall Theorem 4.4.7. With appropriate choices of $b_1, \ldots, b_n$, a 2-bridge knot is reduced and alternating so the diagram realizes the crossing number.

Knots can have arbitrarily high bridge number. This is implicit in the following theorem that was proved by Schubert:

Theorem 4.8.3 (Schubert). *The (p,q)-torus knot has bridge number $\min\{p,q\}$.*

Schubert introduced the bridge number for a specific purpose: to show that a knot can have at most finitely many companions. He first proved vital results on the relative positioning of tori in knot complements. Specifically, he showed that tori are either disjoint or intersect in a controlled way. The following theorem, also due to Schubert, implies that there can be only finitely many disjoint non-isotopic companion tori:

Theorem 4.8.4 (Schubert). *If K is a satellite knot with companion A and pattern of index r, then $b(K) \geq rb(A)$. Furthermore, in the case $r = 1$, when the satellite construction produces the connected sum of knots K_1 and K_2,*

$$b(K_1 \# K_2) = b(K_1) + b(K_2) - 1.$$

In the 1980s Gabai introduced the notion of thin position of a knot. It can be considered either the opposite of bridge position or a refinement thereof. The idea here is to isotope maxima of a knot to lie below minima of a knot whenever possible. See Figure 4.38.

Definition 4.8.5. Let $h : S^3 \to [0,1]$ be a height function. Let $k \subset S^3$ be a representative of a knot K and let $c_1, \ldots, c_n$ be the critical values of $h|_k$ listed in increasing order, i.e., so that $c_1 < \cdots < c_n$. Choose $r_1, \ldots, r_{n-1}$ so that $c_i < r_i < c_{i+1}$. Set $R_i = h^{-1}(r_i)$. The *width of k relative to h*, denoted by $w(k, h)$, is $\sum_i |k \cap R_i|$. The *width of K*, denoted by $w(K)$, is the minimum of this relative width over all representatives of K. We say that K is in *thin position* with respect to h (or merely in thin position if the context is clear) if K is embedded so as to minimize width with respect to h.

The width of the figure 8 knot is 8, as can be seen by its depiction in Figure 4.39. Exhibiting this picture proves that its width can't be more than 8. You will prove in the exercises that only the unknot can have width strictly less than 8.

It is conjectured that for a schematic diagram such as that in Figure 4.38, the braids b_1, b_2, and b_3 can be chosen so that the schematic represents a knot in thin position. If this is indeed the case, then, with appropriate choices of b_1, b_2, and b_3, the knot in Figure 4.38 has width 26.

4.8. Knots and Height Functions

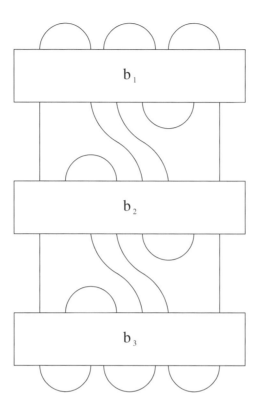

Figure 4.38. *Schematic of a knot in thin position.*

A result relating bridge position and thin position was obtained by Thompson:

Theorem 4.8.6 (Thompson). *If K admits a thin position that is not a bridge position, then the complement of K, $C(K)$, contains a closed essential surface.*

For a proof of this theorem, see [**151**]. The complement of a torus knot contains no closed essential surfaces. This follows from our discussion of essential surfaces in Seifert fibered spaces. It follows that thin position is bridge position for torus knots. Moreover, since the bridge number of torus knots is $\min\{p, q\}$, there are torus knots with arbitrarily large bridge number. Hence there are also torus knots with arbitrarily large width.

Definition 4.8.7. A regular level surface L such that the first critical point of K above L is a maximum and the first critical point of K below L is a minimum is called a *thick level*.

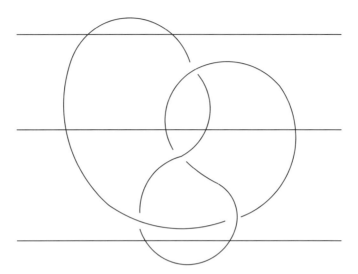

Figure 4.39. *The only regular levels considered in computing the width of the figure 8 knot.*

A regular level surface L such that the first critical point of K above L is a minimum and the first critical point of K below L is a maximum is called a *thin level*.

Example 4.8.8. If K is in bridge position, then the plane separating the maxima from the minima is a thick level and there are no other thick or thin levels.

Given a knot in thin position, we really only need to know the number of intersection points of K with the thick and thin levels in order to compute the width.

Theorem 4.8.9. *Suppose that K is a knot in thin position that intersects the thick levels (in order) $a_0, \ldots, a_m$ times and the thin levels (in order) $b_1, \ldots, b_m$ times. Then*

$$w(K) = \frac{1}{2}\sum_{i=0}^{m} a_i^2 - \frac{1}{2}\sum_{i=1}^{m} b_i^2.$$

Proof. If thin position is bridge position for the knot K, then this result is due to McCrory. In this case $m = 0$, a_0 is twice the bridge number, and we proceed as follows: In computing the width of K, R_1 will intersect K in two points, R_2 will intersect K in four points, and so forth, down to the thick level that intersects K in a_0 points, the regular level below the thick level will intersect K in $a_0 - 2$ points, and so forth, and R_{n-1} will intersect K in two points. Here $|K \cap R_i| = |K \cap R_{i-1}| \pm 2$.

4.8. Knots and Height Functions

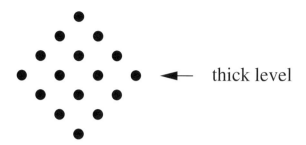

Figure 4.40. *A schematic for computing the width of a knot in bridge position.*

We keep track of one-half the number of points of intersection with the regular levels via a diamond as in Figure 4.40. The top row of the diamond contains one point representing $\frac{1}{2}|K \cap R_1|$. The next row of the diamond contains one more point, representing $\frac{1}{2}|K \cap R_2|$. If the bridge number is l, then there are l maxima. Thus, the thick level will have $2l$ points of intersection with the knot, represented by l dots on the thick level of the diamond. Furthermore, $n = 2l$; hence the corresponding diamond will have l rows. The diamond is equivalent to the $l \times l$ geometric square in the plane with dots on the integer lattice. Thus, there are l^2 dots in the diamond. Hence in this case the width of the knot is $2l^2 = \frac{a_0^2}{2}$.

When thin position is not bridge position, the analogous schematic, rather than being a diamond, becomes more complicated but still shows $\frac{1}{2}w(K)$ dots. See Figure 4.41. The dots in the n rows of the schematic correspond to $\frac{1}{2}|K \cap R_1|, \ldots, \frac{1}{2}|K \cap R_{n-1}|$, respectively. Furthermore, $|K \cap R_i| = |K \cap R_{i-1}| \pm 2$.

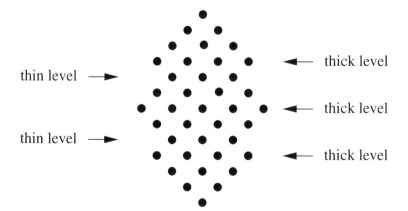

Figure 4.41. *A schematic for computing the width of a knot.*

The numbers of dots in the thick levels of the schematic are $\frac{a_0}{2}, \ldots, \frac{a_m}{2}$ and the numbers of dots in the thin levels of the schematic are $\frac{b_1}{2}, \ldots, \frac{b_m}{2}$. At each thick level t_i for $i = 1, \ldots, n$ in the schematic we can inscribe a "thick" diamond D_i in the schematic whose diagonal is the thick level. See Figure 4.42. Then D_i is equivalent to the $\frac{a_i}{2} \times \frac{a_i}{2}$ geometric square and contains $\frac{a_i^2}{4}$ dots.

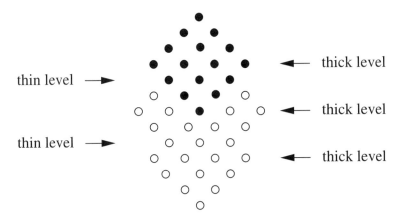

Figure 4.42. *A thick diamond indicated by black dots.*

There will be exactly $m + 1$ thick diamonds inscribed in the schematic. Together, they contain all dots in the schematic. However, the thick diamonds will overlap. So the width of K will be less than $\frac{1}{2} \sum_{i=0}^{n} a_i^2$.

Analogously, we can inscribe m "thin" diamonds $E_1, \ldots, E_m$ in the schematic. Each of these will be equivalent to a $\frac{b_i}{2} \times \frac{b_i}{2}$ geometric square and contain $\frac{b_i^2}{4}$ dots. Now note that the successive thick diamonds D_i, D_{i+1} overlap in the thin diamond E_i. See Figure 4.43.

We leave it as an exercise to show that for each dot d, if d lies in $E_k \cap E_{k+1} \cap \cdots \cap E_l$, then d also lies in $D_{k-1} \cap D_k \cap \cdots \cap D_l$. Thus, for $|D_i|, |E_i|$ the number of dots in D_i, E_i, respectively, the number of dots in the schematic is given by
$$\sum_{i=0}^{m} |D_i| - \sum_{i=1}^{m} |E_i|.$$
The width of the knot is twice the number of dots in the schematic; hence the formula holds. □

In Figure 4.44 we count the dots indicated in black once, we count the dots indicated in white twice and subtract them once, we count the dot indicated in gray three times and subtract it twice.

4.8. Knots and Height Functions

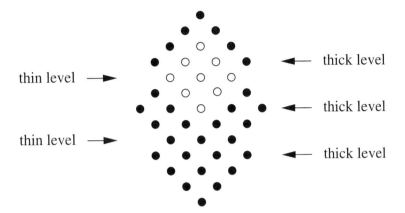

Figure 4.43. *A thin diamond indicated by white dots.*

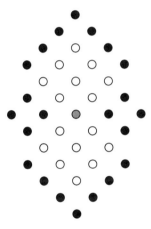

Figure 4.44. *A schematic for computing the width of a knot.*

The behavior of the width under the connected sum of knots is erratic. It is not hard to prove the following proposition:

Proposition 4.8.10.
$$w(K_1 \# K_2) \leq w(K_1) + w(K_2) - 2.$$

Proof. To provide an embedding of $K_1 \# K_2$ of width $w(K_1) + w(K_2) - 2$, we isotope K_1 to lie above K_2. Then we form the connected sum by deleting a small arc containing the minimum of K_1 and a small arc containing the maximum of K_2. See Figure 4.45. □

In some cases, this inequality is sharp.

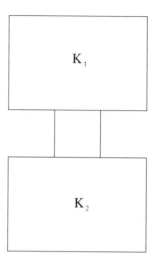

Figure 4.45. *An embedding for $K_1 \# K_2$.*

Definition 4.8.11. A knot is *small* if its complement contains no closed essential surfaces.

Theorem 4.8.12 (Rieck-Sedgwick). *For small knots*

$$w(K_1 \# K_2) = w(K_1) + w(K_2) - 2.$$

The following result, due to Scharlemann and the author, gives us some indication of what to expect:

Theorem 4.8.13 (Scharlemann-Schultens).

$$w(K_1 \# K_2) \geq \max\{w(K_1), w(K_2)\}.$$

The central technique of the proof of this theorem can be used to construct candidates for knots that realize this inequality. This strategy was pursued by Scharlemann and Thompson and was accomplished by Blair and Tomova, who exhibit knots for which this inequality is sharp.

Exercises

Exercise 1. Prove that only the unknot has width strictly less than 8.

Exercise 2. Prove the "easy inequality" in Schubert's Theorem concerning torus knots; specifically, prove that the bridge number of the (p,q)-torus knot is at most $\min\{p,q\}$.

4.9. The Knot Group

Exercise 3. Prove the statement made in the proof of Theorem 4.8.9 that was left as an exercise: "For each dot d, if d lies in $E_k \cap E_{k+1} \cap \cdots \cap E_l$, then d also lies in $D_{k-1} \cap D_k \cap \cdots \cap D_l$."

Exercise 4. Provide an example of a knot that is not small.

4.9. The Knot Group*

To every knot we can associate a group as described below. This can serve to distinguish knots if their associated groups are known to be distinct. In general, distinguishing groups based on their generators and relations can be rather tricky. Nevertheless, the insights of an algebraic point of view are sometimes complementary to the insights from a topological point of view. The knot group and invariants derived from the knot group are an active area of study today.

The knot group is an example of a more general construction: To every connected topological space one can associate the "fundamental group". The knot group is the fundamental group of the complement of the knot. For this reason it is well-defined. We will not cover this fundamental result here. Rather, we will focus on computing knot groups.

The *Wirtinger presentation* provides an algorithm for computing knot groups from knot diagrams:

Step 1. Choose a point far from the knot. See Figure 4.46.

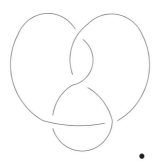

Figure 4.46. *Choosing a point.*

Step 2. To each arc in the knot diagram associate a generator. See Figure 4.47.

Step 3. To each crossing associate a relation as in Figure 4.48, e.g., $x_2 \cdot x_1 \cdot x_2^{-1} \cdot x_4^{-1}$ for the bottom right crossing.

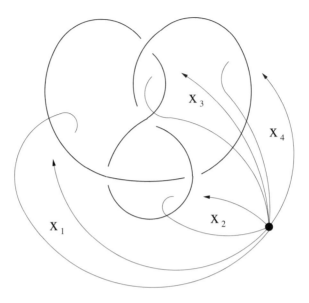

Figure 4.47. *Generators of the knot group.*

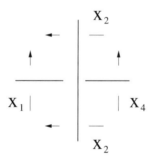

Figure 4.48. *The relation for a crossing: $x_2 x_1 x_2^{-1} x_4^{-1}$.*

Step 4. Simplify if possible.

For the knot group of the figure 8 knot we obtain

$$\langle x_1, x_2, x_3, x_4 \mid x_2 x_1 x_2^{-1} x_4^{-1}, x_1 x_3^{-1} x_1^{-1} x_2^{-1}, x_2 x_3 x_4 x_3^{-1}, x_4 x_3 x_4^{-1} x_1^{-1} \rangle.$$

Substituting

$$x_4 = x_2 x_1 x_2^{-1} \quad \text{and} \quad x_3 = x_4^{-1} x_1 x_4 = x_2 x_1^{-1} x_2^{-1} x_1 x_2 x_1 x_2^{-1}$$

4.10. Covering Spaces

from the first and fourth relations, we obtain

$$\langle x_1, x_2 \mid x_1 x_2 x_1^{-1} x_2^{-1} x_1^{-1} x_2 x_1 x_2^{-1} x_1^{-1} x_2^{-1},$$

$$x_2^2 x_1^{-1} x_2^{-1} x_1 x_2 x_1 x_2^{-1} x_2 x_1 x_2^{-1} x_2 x_1^{-1} x_2^{-1} x_1^{-1} x_2 x_1 x_2^{-1} \rangle$$

$$= \langle x_1, x_2 \mid x_1 x_2 x_1^{-1} x_2^{-1} x_1^{-1} x_2 x_1 x_2^{-1} x_1^{-1} x_2^{-1}, x_2^2 x_1^{-1} x_2^{-1} x_1 x_2 x_1 x_2^{-1} x_1^{-1} x_2 x_1 x_2^{-1} \rangle$$

$$= \langle x_1, x_2 \mid x_1 x_2 x_1^{-1} x_2^{-1} x_1^{-1} x_2 x_1 x_2^{-1} x_1^{-1} x_2^{-1}, x_2 x_1^{-1} x_2^{-1} x_1 x_2 x_1 x_2^{-1} x_1^{-1} x_2 x_1 \rangle$$

$$= \langle x_1, x_2 \mid x_1 x_2 x_1^{-1} x_2^{-1} x_1^{-1} x_2 x_1 x_2^{-1} x_1^{-1} x_2^{-1} \rangle$$

(since in the preceding line the second relation is just a conjugate of the inverse of the first relation).

Exercises

Exercise 1. Calculate the knot group of the trefoil knot. (Hint: The answer should be $\langle a, b | a^2 b^{-3} \rangle$.)

Exercise 2. Calculate the knot group of your favorite knot.

Exercise 3. Prove Dehn's Theorem, which states that a knot is the unknot if and only if the fundamental group of its complement is cyclic. (Hint: Use Exercise 1 from Section 4.3.)

4.10. Covering Spaces*

Recall our discussion of covering spaces in Section 2.4. In the case that X is a manifold, a covering space is just a bundle over X that has a discrete fiber. There is a strong connection between covering spaces and subgroups of the fundamental group:

Theorem 4.10.1. *Let X be a connected topological space. There is a 1-to-1 correspondence between connected covering spaces of X and conjugacy classes of subgroups of $\pi_1(X)$.*

Definition 4.10.2. A cover X' of X is *universal* if it is simply connected.

The universal cover of a space corresponds to the trivial subgroup of the fundamental group of the space. You will prove in the exercises that the universal cover of a manifold is unique up to homeomorphism. In the context of knots, we are particularly interested in the infinite cyclic covering space. It corresponds to the commutator subgroup $[\pi_1(C(K)), \pi_1(C(K))]$ (i.e., the subgroup generated by elements $xyx^{-1}y^{-1}$, for pairs (x, y) of elements of $\pi_1(C(K))$) of the knot group, since $H_1(C(K)) \cong \mathbb{Z}$ and hence $\pi_1(C(K))/[\pi_1(C(K)), \pi_1(C(K))] \cong \mathbb{Z}$).

The infinite cyclic cover of a knot complement is constructed as follows: Let S be a Seifert surface for the knot $K \subset \mathbb{S}^3$. Abusing notation slightly, we also denote $S \cap C(K)$ by S. If we cut $C(K)$ along S, we obtain a manifold M with (connected) boundary. Recall that by Theorem 3.1.14, S is 2-sided. Thus in the boundary of M, we see two remnants of S. We denote one by S^+ and the other by S^-. Now consider a countably infinite collection of copies of M, indexed by $\mathbb{Z}$. Specifically, for each $n \in \mathbb{Z}$, we have M_n such that ∂M_n contains two marked copies of the Seifert surface, S_n^+ and S_n^-. For all n we identify S_n^+ in M_n with S_{n+1}^- in M_{n+1} to obtain M'. See Figure 4.49.

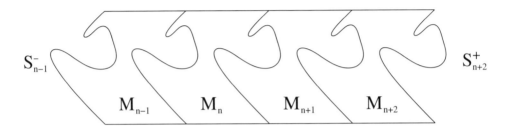

Figure 4.49. *A portion of the infinite cyclic cover of a knot complement.*

There is a natural map $f : M' \to M'$. It takes each point $x \in M_n$ to the corresponding point in M_{n+1}. Let $\langle f \rangle$ be the group generated by f (i.e., the group consisting of powers of the map f). It corresponds to a covering space $(M', C(K), p)$, where $p : M' \to C(K)$ is the quotient map $M' \to M'/\langle f \rangle$.

Here f and powers of f are covering transformations. (The covering transformation f played an implicit role in our computation of the Alexander polynomial.) The infinite cyclic cover factors through k-fold cyclic covers of the form $(M^k = M'/\langle f^k \rangle, C(K), p^k)$, where $p^k : M^k \to C(K)$ is the map induced by p. Thus we obtain a tower of covers:

$$M'$$
$$\downarrow$$
$$M'/\langle f^k \rangle = M^k$$
$$\downarrow p^k$$
$$M'/\langle f \rangle = C(K).$$

Recall also our discussion of branched covering spaces. To obtain the k-fold branched cover of $\mathbb{S}^3$ over K, first construct the k-fold cover $(M^k, C(K), p^k)$ of $C(K)$. Then glue a solid torus V to ∂M^k in such a way that the copies of S meet ∂V in longitudes. We thus obtain a manifold $\hat{M}^k$. Let c be the *core* of V, that is, the simple closed curve given by $\{0\} \times \mathbb{S}^1$. The map q^k given by $z \to z^k$ on the first factor and the identity on the second factor defines a map $V \to V$ such that (V, c, V, c, q^k) is a branched cover. Note that by attaching a solid torus to $C(K)$ in such a way that S meets the solid torus in a longitude, we obtain $\mathbb{S}^3$. Thus by concatenating p^k and q^k to a map r^k we obtain the branched cover $(\hat{M}^k, c, \mathbb{S}^3, K, r^k)$.

One case of special interest is that of a 2-fold branched cover. Note that when k is 2, then a covering transformation is an involution (i.e., its square is the identity). One reason why we are interested in covering spaces and branched covers in particular are the two theorems below:

Theorem 4.10.3 (Alexander). *Every 3-manifold can be obtained as a branched cover over a link in $\mathbb{S}^3$.*

See [6]. For an updated proof, see [40].

In fact, more is true:

Theorem 4.10.4 (Thurston, Hilden-Lozano-Montesinos). *There is a knot K in $\mathbb{S}^3$ (in fact there are many) such that every closed orientable 3-manifold can be realized as a branched cover over K.*

See [64].

Exercises

Exercise 1. Construct at least three distinct 3-manifolds that have a bundle structure with the genus 2 surface as their base space and the circle as their fiber.

Exercise 2. Prove that the universal cover of a manifold is unique up to homeomorphism.

Exercise 3. Consider a 2-bridge knot K and the 2-fold branched cover of $\mathbb{S}^3$ branched over K. $\mathbb{S}^3$ contains sphere submanifolds that separate the 2-bridge knot into two pairs of unknotted arcs. What is the preimage of such a 2-sphere under the covering map? (Hint: Consider Figure 4.50.)

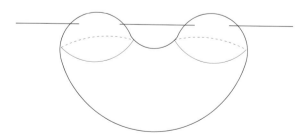

Figure 4.50. *The quotient of a surface by an involution.*

Chapter 5

Triangulated 3-Manifolds

Triangulating a 3-manifold allows us to focus on tetrahedra rather than on the 3-manifold as a whole. For more on the origins of this approach, see [**102**]. Many beautiful results have come out of this theory and there are surely many more to come. Normal surface theory is a sophisticated tool in the theory of 3-manifolds. It was envisioned and developed by Wolfgang Haken who used it to describe algorithms detecting certain properties of 3-manifolds; see [**53**]. This theory is very natural, the sophistication lies in the details. For further reading on normal surface theory, see [**61**] or [**96**].

5.1. Simplicial Complexes

Recall the basic definitions for triangulated manifolds discussed in Section 1.4. We here expand on these definitions for a more substantial discussion of triangulated 3-manifolds. As we will see, the notion of barycentric subdivision interpolates between the more traditional definition of a triangulation of a manifold, where every simplex is required to be embedded and pairs of simplices are allowed to meet in at most one simplex, and the more modern definition used here. The triangulations defined can be subdivided to obtain triangulations in the traditional sense.

In what follows we will consider only locally finite simplicial complexes. The reasoning used in Lemma 1.4.28 then allows us to view our simplicial complexes as quotient spaces of standard simplices.

Definition 5.1.1. Let K be a k-simplex. Then for $r \leq k$, the *r-skeleton* K^r of K is the collection $K^r = \{[s] \in K; \dim[s] \leq r\}$.

Example 5.1.2. Figure 5.1 depicts the 1-skeleton of the simplicial complex depicted in Figure 1.18.

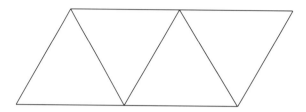

Figure 5.1. *The 1-skeleton of a simplicial complex.*

Remark 5.1.3. The r-skeleton of a simplicial complex K is a subcomplex of K.

Definition 5.1.4. For a point v in the standard k-simplex $[s] = [v_0, \ldots, v_k]$, there is a vector $(a_0, \ldots, a_k)$ with $a_i \in [0, 1]$ such that $v = a_0 v_0 + \cdots + a_k v_k$. The $(k+1)$-tuple $(a_0, \ldots, a_k)$ is called the *barycentric coordinates* of v.

Remark 5.1.5. Elementary linear algebra shows that the barycentric coordinates of a point $v \in [s]$ are unique.

In $\mathbb{R}^{k+1}$, most collections $w_0, \ldots, w_k$ of $k+1$ points span a k-simplex $[w_0, \ldots, w_k]$. Note that there will always be an affine map from the standard k-simplex to $[w_0, \ldots, w_k]$. Below, we define a 0-simplex called the barycenter of a simplex and a collection of simplices called the barycentric subdivision of a simplex.

Definition 5.1.6. The point $\frac{1}{k+1} v_0 + \cdots + \frac{1}{k+1} v_k$ is called the *barycenter* of $[v_0, \ldots, v_k]$ and is denoted by $b([v_0, \ldots, v_k])$.

Remark 5.1.7. Note that $b([v_0]) = v_0$.

Definition 5.1.8. Let K be a simplicial complex. Define a *partial ordering* on K by $[s_1] \leq [s_2]$ if and only if $[s_1]$ is a face of $[s_2]$. We write $[s_1] < [s_2]$ when $[s_1] \leq [s_2]$ and $[s_1] \neq [s_2]$.

Below we define the barycentric subdivision of a simplicial complex K. Subdivisions can be defined more generally, but we will not do so here. The idea is to take each standard simplex in the simplicial complex and subdivide it into smaller simplices. This can be achieved by adding the barycenters of edges, thereby subdividing each edge into two edges, then adding the barycenters of 2-dimensional faces and adding six edges in each face between the barycenter and the vertices of these (subdivided) edges of the 2-dimensional face and proceeding inductively adding simplices spanned by barycenters to obtain a simplicial complex.

5.1. Simplicial Complexes

Definition 5.1.9. Let $[s]$ be the standard k-simplex. Let $[s_0], \ldots, [s_l]$ be faces of $[s]$ such that $[s_0] < [s_1] < \cdots < [s_l]$. Consider the simplex $[b([s_0]), b([s_1]), \ldots, b([s_l])]$. The *first barycentric subdivision* of $[s]$ is the simplicial complex consisting of all simplices of this form:

$$\{[b([s_0]), b([s_1]), \ldots, b([s_l])] :$$
$$[s_0], [s_1], \ldots, [s_l] \text{ faces of } [s] \text{ such that } [s_0] < [s_1] < \cdots < [s_l]\}.$$

Let K be a simplicial complex. Then $|K|$ is the quotient space obtained by identifying standard simplices. The *first barycentric subdivision* of K, denoted by $K^{(1)}$, is the simplicial complex consisting of the union of first barycentric subdivisions of the standard simplices in K.

Abusing terminology, we continue to think of $K^{(1)}$ as a quotient space obtained from standard simplices.

Definition 5.1.10. The simplicial complex $K^{(n)} = (((K^{(1)})^{(1)}) \ldots)^{(1)}$ is called the *n*-th barycentric subdivision of K.

Caution: $K^n \neq K^{(n)}$.

Example 5.1.11. Figure 5.2 depicts the first barycentric subdivision of the simplicial complex depicted in Figure 1.18.

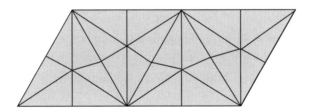

Figure 5.2. *The first barycentric subdivision of the simplicial complex in Figure* 1.18.

Recall that the modern definition of triangulation allows a simplex in a manifold to share more than one face with another simplex. It also allows distinct faces of a simplex to coincide. You will prove in the exercises that the second barycentric subdivision of a modern triangulation is a triangulation in the traditional sense: Each simplex is embedded and any two simplices share at most one face.

In studying simplicial complexes, it is of interest to understand the local picture near a vertex. The local picture of a simplicial complex near a vertex is captured by the following concept:

Definition 5.1.12. The *star* of v, for v a vertex of a simplicial complex K, is the set
$$St(v) = \{(s) \in K : v \in [s]\}.$$

More generally, the star of a subcomplex K' of K is the set
$$St(K') = \{(s) \in K : K' \cap [s] \neq \emptyset\}.$$

Remark 5.1.13. Here $St(v)$ and $St(K')$ are not themselves simplicial complexes since they are unions of open simplices. We will nevertheless denote the set of points in $St(v)$ by $|St(v)|$.

Recall the definition of a simplicial map in Section 1.4. A related notion is the following:

Definition 5.1.14. Let K, L be simplicial complexes and let $f : |K| \to |L|$ be a continuous map. A simplicial map $\phi : |K| \to |L|$ is a *simplicial approximation* to f if $f(|St(v)|) \subset |St(\phi(v))|$ for each vertex v of K.

Theorem 5.1.15. *Let K, L be finite simplicial complexes and let $f : |K| \to |L|$ be a continuous map. For sufficiently large n, there exists a simplicial map $\phi : |K^{(n)}| \to |L|$ that is a simplicial approximation of f.*

A proof of this theorem can be found in [148]. The definitions there differ slightly from the definitions used here. However, with appropriate modifications, the proof still carries over.

Example 5.1.16. Consider the continuous map of a 1-dimensional simplicial complex into a 2-dimensional simplicial complex pictured in Figure 5.3. To construct a simplicial approximation, we first need to consider the condition on stars. Note that there is no vertex in the 2-dimensional simplicial complex that could be the image of the vertex v_3 of the 1-dimensional simplicial complex and satisfy the star condition. Thus we must first take the barycentric subdivision of the 1-dimensional simplicial complex to construct a simplicial approximation. See Figure 5.4. The simplicial approximation is pictured in Figure 5.5.

Definition 5.1.17. Let K be a simplicial complex. For $v \in K^0$, consider the star $St(v; K^{(1)})$ of v in the first barycentric subdivision of K. Take
$$\overline{St(v)} = \{[s] : (s) \in St(v)\}.$$
The *link* of v, denoted by $\mathrm{link}(v)$, is defined by
$$\mathrm{link}(v) = \overline{St(v; K^{(1)})} \backslash St(v; K^{(1)}).$$

5.1. Simplicial Complexes

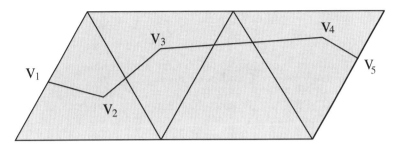

Figure 5.3. *A continuous map of a 1-dimensional simplicial complex into a 2-dimensional simplicial complex.*

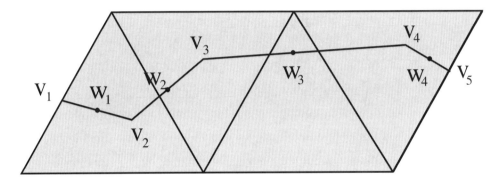

Figure 5.4. *The barycentric subdivision.*

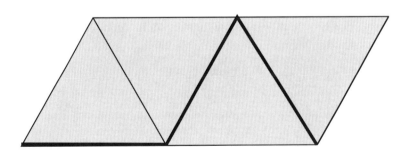

Figure 5.5. *A simplicial approximation.*

The subject of general position is a rather intricate one and is briefly treated (in the DIFF category) in Appendix A. We require a special case of general position in the TRIANG category that we introduce here.

Definition 5.1.18. Let v be a point in $\mathbb{R}^n$ and let A be a subset of $\mathbb{R}^n$. The pair (v, A) is in *general position* if $v \notin A$ and, for each $a_1, a_2 \in A$ with $a_1 \neq a_2$, $[v, a_1] \cap [v, a_2] = \{v\}$.

Definition 5.1.19. Let (v, A) be in general position. The set

$$\bigcup_{a \in A} [v, a]$$

is called the *cone* of v over A and is denoted by $v * A$.

Example 5.1.20. Figure 5.6 depicts the cone of a pair consisting of a point v and a line segment A in $\mathbb{R}^2$.

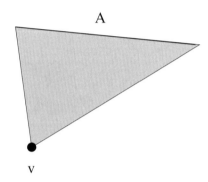

Figure 5.6. *The cone of the pair (v, A) where A is a line segment in $\mathbb{R}^2$.*

Exercises

Exercise 1. Let (M, K) be a triangulated manifold. Prove that the second barycentric subdivision of K is a triangulation of M in the traditional sense.

Exercise 2. Let $[s] = [v_0, \ldots, v_k]$ be the standard k-simplex. Prove that $(v_0, [v_1, \ldots, v_k])$ is in general position and $v_0 * [v_1, \ldots, v_k] = [s]$.

Exercise 3. Prove that for a simplicial complex K, the collection $\{St(v)\}_{v \in K^0}$ is a covering of $|K|$.

Exercise 4. Let $f : |K| \to |L|$ be a continuous map and let $\phi : |K| \to |L|$ be a simplicial approximation. Prove that $\phi : |K| \to |L|$ is homotopic to f (via a "straight line homotopy").

5.2. Normal Surfaces

Normal surfaces, defined below, are a natural class of surfaces to consider in the context of triangulated 3-manifolds. We shall see that incompressible surfaces can be isotoped to be normal surfaces. In this context, neither incompressible surfaces nor normal surfaces are subcomplexes of the triangulations of the 3-manifolds under consideration. Nevertheless, via the tool

5.2. Normal Surfaces

of normal surfaces, the triangulation of the manifold helps us understand the incompressible surface.

Definition 5.2.1. A (properly embedded) arc on a 2-dimensional face $[f]$ of a 3-simplex $[s]$ is a *normal arc* if its endpoints lie on distinct edges of $[f]$. A simple closed curve c on the boundary of a 3-simplex $[s]$ is a *normal curve* if every component of intersection of c with a 2-dimensional face $[f]$ of $[s]$ is a normal arc. See Figures 5.7 and 5.8

Normal arcs lie on the 2-dimensional faces of a 3-simplex $[s]$. Here $[s]$ inherits a geometry. This allows us, for simplicity, to assume that normal arcs are straight line segments.

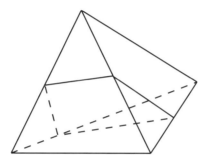

Figure 5.7. *A normal curve.*

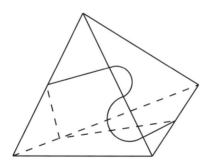

Figure 5.8. *A curve that is not normal.*

Definition 5.2.2. The *length* of a normal curve c on the boundary of a 3-simplex $[s]$ is the number of points in $c \cap |[s]^1|$.

Example 5.2.3. The length of the normal curve depicted in Figure 5.7 is 4.

Lemma 5.2.4. *A closed normal curve on the boundary of a 3-simplex either has length* 3 *or* 4 *or it meets some edge more than once.*

Proof. Let $[s]$ be a 3-simplex and let c be a closed normal curve on $\partial|[s]|$. Here $\partial|[s]^2|$ is homeomorphic to a sphere. Thus c is a Jordan curve on the sphere. It separates $\partial|[s]^2|$ into an "inside" (disk) and an "outside" (disk). Consider $|[s]^0|$. It consists of four vertices. These vertices are distributed among the inside and outside disks.

Claim. Each disk contains at least one vertex.

Suppose a disk contains no vertices. How do the edges of $[s]$ intersect this disk? They intersect it in arcs. An outermost arc cuts off a subarc of c that has both endpoints on the same edge. This implies that c is not normal, a contradiction. Thus the disk is disjoint from $|[s]^1|$ and is contained in a 2-dimensional face. This is also a contradiction, as a 2-dimensional face contains no closed normal curve.

It follows that, up to renaming, there are only two possibilities: The "inside" of c contains either one or two points of $|[s]^0|$. Suppose that the "inside" of c contains only the vertex v. Let e_1, e_2, e_3 be the edges incident to v. Then c must intersect e_1, e_2, e_3 because the other vertices to which e_1, e_2, e_3 are incident lie "outside" of c. In fact, c must intersect each of e_1, e_2, e_3 an odd number of times. Similarly, it must meet the other three edges an even number of times. Thus if c meets no edge more than once, then it has length 3.

Suppose now that the "inside" of c contains the vertices v_1, v_2. Then by analogous reasoning, there are four edges that are met an odd number of times and two edges that are met an even number of times. If the former edges each meet c once and the latter are disjoint from c, then c has length 4. Otherwise, c meets some edge more than once. $\square$

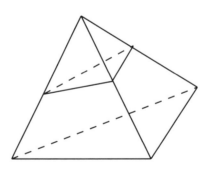

Figure 5.9. *A normal curve of length 3.*

Definition 5.2.5. A *normal triangle* in a 3-simplex $[s]$ is a (properly embedded) disk whose boundary is a normal curve of length 3. A *normal*

5.2. Normal Surfaces

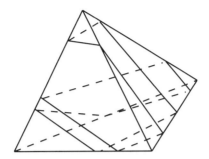

Figure 5.10. *A normal curve of length* 12.

quadrilateral in a 3-simplex $[s]$ is a (properly embedded) disk whose boundary is a normal curve of length 4. A *normal disk* is a normal triangle or quadrilateral.

Example 5.2.6. The triangle spanned by the normal curve of length 3 in Figure 5.9 is a normal triangle. The quadrilateral spanned by the normal curve of length 4 in Figure 5.7 is a normal quadrilateral.

Non-Example. The normal curve in Figure 5.10 does not span a normal disk.

Definition 5.2.7. Let (M, K) be a triangulated 3-manifold. A *normal surface* in (M, K) (or simply in M when it is clear what triangulation is meant) is a surface $S \subset M$ such that for every 3-simplex $[s]$ in K, $S \cap |[s]|$ consists of disjoint normal disks in $[s]$.

Example 5.2.8. Recall that the 3-sphere can be obtained by identifying the faces of two 3-simplices as in Figure 5.11. The two shaded 2-simplices in Figure 5.11 identify to a normal 2-sphere.

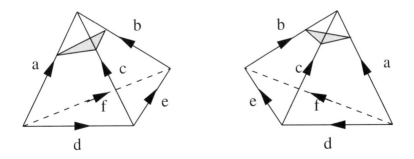

Figure 5.11. *A normal 2-sphere in the 3-sphere obtained by identifications.*

Example 5.2.9. The 3-torus is obtained by identifying eight appropriately chosen reflections of the cube pictured in Figure 5.12. A normal 2-torus in this triangulated 3-torus is obtained by choosing four of these to contain reflections of the square pictured.

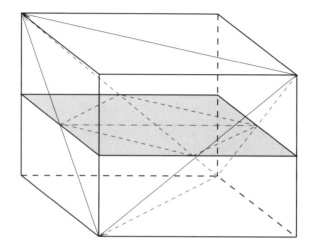

Figure 5.12. *A portion of a normal 2-torus in the 3-torus.*

Definition 5.2.10. The *weight* of a surface S in a triangulated 3-manifold (M, K) is the number of components of $S \cap |K^1|$. It is denoted by $w(S)$. Similarly, the *measure* of S, $m(S)$, is the number of components of $S \cap (|K^2| \backslash |K^1|)$.

Example 5.2.11. The weight of the normal 2-sphere in the 3-sphere pictured in Figure 5.11 is 3. Its measure is also 3.

Example 5.2.12. The weight of the normal 2-torus in Example 5.2.9 is 12. Its measure is 36.

In the context of normal surfaces we often work with the following, more restricted, notion of isotopy.

Definition 5.2.13. A *normal isotopy* is an isotopy through normal surfaces.

Theorem 5.2.14. *Let M be a closed irreducible 3-manifold containing an incompressible surface S. Then for any triangulation (M, K) of M there is an isotopy that takes S to a normal surface in (M, K).*

The analogous result holds if M is merely compact but S is both incompressible and boundary incompressible.

5.2. Normal Surfaces

Proof. Let (M, K) be a triangulation of M and let S be an incompressible surface in M. Isotope S so that $(w(S), m(S))$ is minimal (in the dictionary order). (We assume that S is in general position with respect to the triangulation. In particular, S is disjoint from vertices, intersects edges in a finite number of points, and intersects 2-simplices in a finite number of arcs.)

Let $[s]$ be a 3-simplex of K and let $[f]$ be a 2-dimensional face of $[s]$. Since M is irreducible and S is incompressible, a standard innermost disk argument (see Section 3.9) shows that simple closed curves in $S \cap |[f]|$ can be removed. Removing a simple closed curve lowers $(w(S), m(S))$ by $(0, 1)$. Minimality of $(w(S), m(S))$ thus ensures that $S \cap |[f]|$ contains no closed components. (See Figures 5.13 and 5.14.)

A standard outermost arc argument (see Section 3.9) shows that simple arcs in $S \cap |[f]|$ with both endpoints on the same edge can be removed. Removing such an arc lowers $(w(S), m(S))$ by $(2, 1)$. Thus minimality of $(w(S), m(S))$ ensures that, after normal isotopy, $S \cap |[f]|$ consists of normal arcs. Since this is true for all 2-dimensional faces of $[s]$, $S \cap \partial|[s]|$ consists of normal curves.

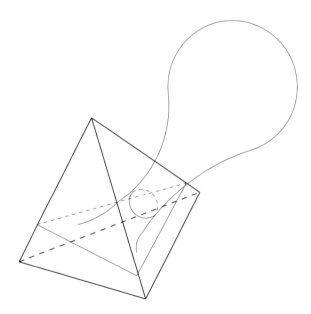

Figure 5.13. *A closed component of intersection of S with K^2.*

Let $\tilde{c}$ be a normal curve in $S \cap \partial|[s]|$. We need to show that $\tilde{c}$ bounds a disk in $|[s]|$. Since $\tilde{c}$ bounds a disk E in the 3-ball $|[s]|$ and since S is incompressible, $\tilde{c}$ must in fact bound a disk $\tilde{S}$ in S. A priori the interior of E may not be disjoint from S, but a standard innermost disk argument shows how to eliminate simple closed curves of intersection in $E \cap S$. Thus

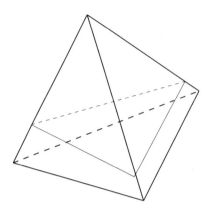

Figure 5.14. *After the isotopy.*

$E \cup \tilde{S}$ is a 2-sphere in an irreducible 3-manifold and hence bounds a 3-ball B. If $\tilde{S}$ does not lie entirely in $|[s]|$, then either $w(\text{int}(\tilde{S}))$ or $m(\text{int}(\tilde{S}))$ or both are positive and hence B describes an isotopy lowering $(w(S), m(S))$ by this amount, a contradiction. Thus $\tilde{c}$ bounds the disk $\tilde{S} = S \cap |[s]|$.

We now need to show that $\tilde{S}$ is a normal disk. By Lemma 5.2.4 it suffices to show that $\partial \tilde{S}$ does not meet any edges of $[s]$ more than once. Suppose that $\partial \tilde{S}$ meets the edge $[e]$ of $[s]$ more than once. Here $\tilde{c} = \partial \tilde{S}$ partitions $\partial|[s]|$ into two disks, D, D'. One of these disks, say D, meets (e) in a subarc α connecting two adjacent points of intersection in $S \cap |[e]|$. Furthermore, $\tilde{S}$ is isotopic to D. This guarantees the existence of an arc β in the interior of $\tilde{S}$ and a disk E with $\partial E = \alpha \cup \beta$ and interior in (s). The disk E can be used to describe an isotopy analogous to the isotopy used in the standard outermost arc argument. This isotopy lowers $w(S)$ by 2. Thus the minimality of $(w(S), m(S))$ guarantees that $\partial \tilde{S}$ does not meet any edges of $[s]$ more than once and hence, by Lemma 5.2.4, $\tilde{S}$ is a normal disk.

Applying this reasoning to all 3-simplices in K shows that S is a normal surface. □

Exercises

Exercise 1. Choose a triangulation $(\mathbb{S}^3, K)$ of $\mathbb{S}^3$ and give an example of a normal surface in $(\mathbb{S}^3, K)$.

Exercise 2. Is the converse of Theorem 5.2.14 true? (I.e., is a normal surface in a triangulated 3-manifold necessarily incompressible?)

5.3. Diophantine Systems

One advantage of working with normal surfaces lies in the fact that local behavior is controlled. Given a normal surface S in a triangulated 3-manifold (M, K), consider a 3-simplex $[s]$ in K. There are exactly four normal isotopy types of normal triangles in $[s]$: Indeed, a normal curve c of length 3 partitions the vertices of $[s]$ into one vertex on one side of c and the remaining three vertices on the other side of c. Thus, up to isotopies through normal curves of length 3, there are exactly four such curves. Moreover, a normal triangle is determined by its boundary. Likewise, you will prove in the exercises that there are exactly three normal isotopy types of normal quadrilaterals in $[s]$ and that representatives of distinct types must intersect. Thus $S \cap [s]$ determines a vector with seven entries, $(t^1, t^2, t^3, t^4, q^1, q^2, q^3)$. The first four entries correspond to the number of normal triangles of a given normal isotopy type, the last three entries correspond to the number of normal quadrilaterals of a given normal isotopy type, and only one of the latter three entries can be non-zero. A vector of this form (seven non-negative integral entries, with only one of the last three entries non-zero) is said to *satisfy the square restriction*. Thus if t is the number of 3-simplices in a triangulated 3-manifold (M, K), then S determines a vector with $7t$ entries that satisfies the square restriction.

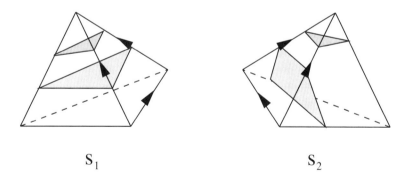

Figure 5.15. *A normal surface in adjacent 3-simplices.*

Now consider two adjacent 3-simplices $[s_1]$ and $[s_2]$ (see Figure 5.15) with vectors $(t_1^1, t_1^2, t_1^3, t_1^4, q_1^1, q_1^2, q_1^3)$, $(t_2^1, t_2^2, t_2^3, t_2^4, q_2^1, q_2^2, q_2^3)$. The normal triangles and normal quadrilaterals in $[s_1]$ and $[s_2]$ must match up along any 2-simplices in which $[s_1]$ and $[s_2]$ meet. This places restrictions on the nature and number of normal triangles and quadrilaterals that can occur in $[s_1]$ and $[s_2]$. To be precise, there are three normal isotopy types of normal arcs in each 2-simplex. For each such normal arc, there is one normal isotopy type of normal triangle and one normal isotopy type of normal quadrilateral that

give rise to this normal arc. Thus there are three equations of the form
$$t_1^i + q_1^j = t_2^k + q_2^l$$
associated with a 2-simplex along which $[s_1]$ and $[s_2]$ meet. Hence there are $6t$ *gluing equations* that must be satisfied by the vector determined by S.

We wish to use linear algebra to help us find normal surfaces in a triangulated 3-manifold. To this end we need to make sense out of the sum of two normal surfaces. More specifically, given two normal surfaces S_1, S_2 and corresponding vectors $\vec{v}_1, \vec{v}_2$, we wish to define the sum $S_1 + S_2$ corresponding to $\vec{v}_1 + \vec{v}_2$. This is possible, but we must proceed with caution: In a 3-simplex, the normal triangles and quadrilaterals corresponding to S_1 can intersect the normal triangles and quadrilaterals corresponding to S_2. In the 3-simplex, we may be able to isotope these apart, but note that the surface connects up to other normal disks in adjacent 3-simplices. The isotopy might not extend beyond the 3-simplex to make progress globally. This will certainly be the case if S_1 and S_2 can't be made disjoint in M.

Figure 5.16. *A regular switch.*

Figure 5.17. *An irregular switch.*

The operation in Figure 5.16 is called a *regular switch*. (Compare this with the operation in Figure 5.17 that is not a regular switch.) If, in a given 3-simplex $[s]$, two normal triangles intersect, then the regular switches on the 2-dimensional faces of $[s]$ extend into (s) to define a cut and paste operation on the normal triangles resulting in two disjoint normal triangles of the same normal isotopy types as the original pair. Similarly, if a normal triangle and

5.3. Diophantine Systems

a normal quadrilateral intersect in $[s]$, then the regular switches on the 2-dimensional faces of $[s]$ extend into (s) to define a cut and paste operation resulting in a disjoint normal triangle and a normal quadrilateral of the same normal isotopy types as the original pair. If two normal quadrilaterals intersect in $[s]$, the regular switches on the 2-dimensional faces of $[s]$ extend into (s) only if the two quadrilaterals are of the same normal isotopy type. If the two quadrilaterals are of different normal isotopy types, the regular switches on the 2-dimensional faces of $[s]$ do not extend into (s) (the cut and paste operation is not defined). We leave this as an exercise for the reader. Two normal surfaces S_1, S_2 in the triangulated 3-manifold M are said to satisfy the *square restriction* if for every 3-simplex $[s]$ in M, at most one normal isotopy type of quadrilateral occurs in $(S_1 \cup S_2) \cap |[s]|$. Note that the surfaces in question need not be connected.

Definition 5.3.1. Let S_1, S_2 be a pair of normal surfaces in the 3-manifold M that satisfy the square restriction. We denote the result of performing the cut and paste operations discussed above by $S_1 + S_2$.

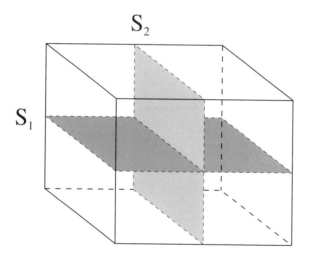

Figure 5.18. The surfaces S_1 and S_2.

Given a normal surface S, the normal triangles and quadrilaterals that constitute S match up along the faces of adjacent 3-simplices. Thus the vector $\vec{v}$ determined by S satisfies the square restriction and the $6t$ gluing equations; i.e., $A\vec{v} = 0$ for some matrix A with non-negative integer entries.

Conversely, given a vector with $7t$ non-negative integral entries that satisfy the gluing equations and the square restriction, we can build a normal surface from the corresponding number of normal triangles and quadrilaterals. Note that, up to normal isotopy, there is only one way that a given

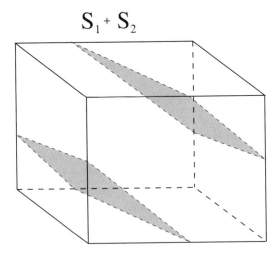

Figure 5.19. *The surface $S_1 + S_2$.*

collection of normal triangles and normal quadrilaterals satisfying the square restriction can be embedded in a 3-simplex. See Figure 5.20.

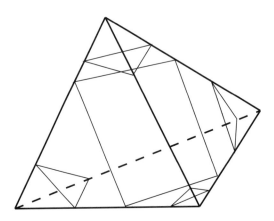

Figure 5.20. $S_i \cap [s]$.

The normal triangles and quadrilaterals match up to form a surface because the gluing equations are satisfied. Furthermore, consider two surfaces, S_1, S_2 corresponding to the same vector. In each simplex $[s]$, $S_1 \cap |[s]|$ must be properly isotopic to $S_2 \cap |[s]|$. This isotopy can be realized, for instance, by taking the straight line homotopy on the edges of $[s]$ (which intersect S_1 and S_2 in ordered sets of points of the same cardinality) and extending across the faces and the interior of $[s]$. Moreover, these normal isotopies in the simplices piece together to a normal isotopy between S_1 and S_2. Hence

5.3. Diophantine Systems

there is a 1-to-1 correspondence between normal surfaces in a triangulated 3-manifold, considered up to normal isotopy, and non-negative integral solutions to the system of equations described above. By considering the non-negative solutions of a system of equations we are really considering a system of equations and inequalities. A system of linear equations and inequalities with integral coefficients is called a *Diophantine system*.

Diophantine systems have been studied extensively. We wish to describe their solution sets to further our understanding of normal surfaces. The following theorem is our key to understanding. It gives us a finite generating set for all our solutions. This provides a finite set of normal surfaces that generate all of our normal surfaces up to normal isotopy!

Theorem 5.3.2. *Let $A\mathbf{x} = 0, x_i \geq 0 \; \forall i$ be a Diophantine system. There is a finite set of integral solutions that generates the full set of integral solutions to the Diophantine system under the $+$ operation.*

Proof. Let A be an $(m \times n)$-matrix with non-negative integer coefficients. We consider $\mathbb{Z}^n \subset \mathbb{R}^n$. Set
$$X = \{\mathbf{x} = (x_1, \ldots, x_n) \in \mathbb{R}^n \; : \; x_i \geq 0 \text{ and } A\mathbf{x} = 0\}.$$
Note that each of the rows of A defines a hyperplane through the origin. Thus X lies in the intersection of these m hyperplanes. The intersection of these hyperplanes is a subspace V of $\mathbb{R}^n$ of dimension d at least $n - m$. So X is the intersection of this subspace with the first quadrant of $\mathbb{R}^n$. In particular, X is convex. Now consider the hyperplane H defined by
$$x_1 + \cdots + x_n = 1.$$
Then $H \cap X$ is a polyhedron that is the convex hull of a finite set of points C, where C consists of the vertices of $H \cap X$.

Let $y \in C$. It is not hard to see (by considering Gauss-Jordan elimination) that the entries in y are rational. Thus if we multiply y by the least common multiple of the denominators of its entries, we obtain an integral point x. Let C' be the collection of integral points thus obtained. Note that C' represents a set of vectors $w_1, \ldots, w_l$ such that every vector in X can be written as $\sum_{i=1}^{l} a_i w_i$ with $a_i \geq 0 \; \forall i$.

Consider the "parallelogram"
$$P' = \left\{ \sum_{x \in C'} t_x x \, | \, t_x \in [0, 1] \right\}.$$
Then P' is compact. Thus the set $L = P' \cap \mathbb{Z}^n$ is a finite number of points. Here L is a finite set of non-negative solutions of $A\mathbf{x} = 0$. It remains to show that L generates the full set of non-negative integral solutions of $A\mathbf{x} = 0$ under the $+$ operation.

Let $z \in X \cap \mathbb{Z}^n$. Then $z = a_1 w_1 + \cdots + a_l w_l$ for some $a_1, \ldots, a_l \in \mathbb{R}_+$. But then

$$z - (\lfloor a_1 \rfloor w_1 + \cdots + \lfloor a_l \rfloor w_l) \in P',$$

where $\lfloor a \rfloor$ denotes the greatest lesser integer of a. Furthermore, all entries of the vector $z - (\lfloor a_1 \rfloor w_1 + \cdots + \lfloor a_l \rfloor w_l)$ are differences of integers, hence themselves integers. In particular, $z - (\lfloor a_1 \rfloor w_1 + \cdots + \lfloor a_l \rfloor w_l) \in L$. Now

$$z = (z - (\lfloor a_1 \rfloor w_1 + \cdots + \lfloor a_l \rfloor w_l)) + (\lfloor a_1 \rfloor w_1 + \cdots + \lfloor a_l \rfloor w_l).$$

Since $\lfloor a_1 \rfloor w_1 + \cdots + \lfloor a_l \rfloor w_l$ is generated by L under the $+$ operation, $X \cap \mathbb{Z}^n$ is generated by L under the $+$ operation. $\square$

Definition 5.3.3. A *set of fundamental solutions* to a Diophantine system is a finite set of solutions $\mathcal{F}$ such that any solution of the system is a linear combination of elements of $\mathcal{F}$ with positive integral coefficients and such that no element of $\mathcal{F}$ is a positive integral linear combination of other elements of $\mathcal{F}$. The elements of $\mathcal{F}$ are called *fundamental solutions*.

Remark 5.3.4. A set of fundamental solutions can be chosen from the finite set L described in Theorem 5.3.2 by eliminating redundancies.

Given a 3-manifold M, we can triangulate M and write down the corresponding Diophantine system. Theorem 5.3.2 provides a finite set of solutions that yields a set of fundamental solutions after eliminating redundancies. Each fundamental solution corresponds to a *fundamental surface*. The (finitely many) fundamental surfaces generate all normal surfaces. In other words, every normal surface is a finite sum of fundamental surfaces under the $+$ operation. In the exercises you will show that a normal surface is a fundamental surface if and only if it is not the sum of two (non-empty) normal surfaces. Note that a fundamental surface is thus necessarily connected.

Many properties of surfaces can be traced back to the analogous properties for the fundamental surfaces of which they are the sum. See, for instance, the results below.

Lemma 5.3.5. *Suppose that F is a connected normal surface (considered up to normal isotopy) in the 3-manifold M. Suppose further that $F = G + H$, for G, H normal surfaces in M, and that G, H are chosen (up to normal isotopy) so that the number of components of $G \cap H$ is minimal among all pairs of summands of F. Then G and H are connected.*

Proof. Suppose that $H = H_1 \sqcup H_2$ for $H_1, H_2 \neq \emptyset$. Set $G' = G + H_1$. Then $F = G + H = G + H_1 + H_2 = G' + H_2$. Here the components of $G \cap H$ consist of the components of $G \cap H_1$ together with the components of $G \cap H_2$. Since F is connected, neither of the latter two sets are empty. Thus the number of components of $G' \cap H_2$ is less than the number of components of $G \cap H$. But this contradicts the minimality of the number of components of $G \cap H$ among all pairs of summands of F. Thus G and H are connected. □

How incompressibility features into this equation is a more difficult question, but it was answered by Jaco and Oertel (see [**68**]):

Theorem 5.3.6 (Jaco-Oertel). *Suppose that S is a 2-sided incompressible normal surface that is positioned so that $w(S)$ is minimal within the isotopy class of S. Suppose further that $S = S_1 + S_2$. Then both S_1 and S_2 are incompressible.*

Often we are interested in Haken manifolds and hence whether or not a 3-manifold contains an incompressible surface. It follows from Jaco and Oertel's Theorem that a 3-manifold contains an incompressible surface if and only if it contains an incompressible fundamental surface. In the next chapter we will see how incompressibility can be expressed algebraically. Existence of an incompressible fundamental surface can then be checked algorithmically.

Corollary 5.3.7 (Haken). *There is an algorithm to detect whether or not a 3-manifold contains an incompressible surface.*

Exercises

Exercise 1. Prove that there are exactly three types of normal quadrilaterals in a 3-simplex. Prove also that any two quadrilaterals of distinct types must intersect.

Exercise 2. Show that the + operation on normal surfaces is commutative and associative.

Exercise 3. Show that for normal surfaces P, Q, the Euler characteristic is additive:
$$\chi(P+Q) = \chi(P) + \chi(Q).$$

Exercise 4. Show that a normal surface is fundamental if and only if it is not the sum of two (non-empty) normal surfaces.

Exercise 5*. Let Q_1, Q_2 be normal quadrilaterals in the 3-simplex $[s]$. Show that regular switches on the 2-dimensional faces of $[s]$ extend to a cut and paste operation on $Q_1 \cup Q_2$ if and only if Q_1 and Q_2 are of the same normal type.

5.4. 2-Spheres*

One application of normal surface theory lies in identifying 2-spheres in 3-manifolds. Understanding 2-spheres in 3-manifolds provides a foundation for understanding prime decompositions of 3-manifolds. We here discuss a result of Kneser, proved in 1929, and a result of Haken that is obtained by an analogous argument. Combined, the two theorems are known as Kneser-Haken finiteness. These results are interesting in their own right. As we will see, Kneser's result is also a crucial step in obtaining prime decompositions.

Definition 5.4.1. Let M be a 3-manifold. A *punctured* M is a 3-manifold homeomorphic to $M \backslash (\text{finite union of pairwise disjoint 3-balls})$.

Let $S = S_1 \sqcup \cdots \sqcup S_k$ be a disjoint union of 2-spheres in a 3-manifold M. We say that S is an *independent* set of 2-spheres if no component of $M \backslash S$ is a punctured 3-sphere.

Example 5.4.2. Let $M = M_1 \# \ldots \# M_k$, where $M_i \neq S^3$ for all i. Then M contains an independent set $S = S_1 \sqcup \cdots \sqcup S_{k-1}$ of $k-1$ 2-spheres.

Note that a 3-manifold can contain any number of 2-spheres. Simply look at a regular neighborhood of a point. Its boundary is a 2-sphere. Such 2-spheres and other inessential 2-spheres are not interesting in this context. Another type of 2-sphere that is uninteresting in this context is the following: Let S_1, S_2 be disjoint essential 2-spheres in a connected 3-manifold. Since the 3-manifold is path-connected, there is a simple arc α connecting S_1 and S_2. The boundary of a regular neighborhood of $S_1 \cup S_2 \cup \alpha$ consists of three 2-spheres, one parallel to S_1, one parallel to S_2, and one additional 2-sphere, $\tilde{S}$. If S_1 and S_2 are essential and non-parallel, then $\tilde{S}$ is also an essential 2-sphere; it is not parallel to S_1 or S_2, but $S_1 \sqcup S_2 \sqcup \tilde{S}$ is not independent! In 1929, Kneser shed light on how many independent 2-spheres can sit in a 3-manifold (see [80]).

Theorem 5.4.3 (Kneser's Theorem). *Let (M, K) be a triangulated 3-manifold containing no punctured $\mathbb{R}P^3$. Suppose that the number of 3-simplices in K is t. If M contains an independent set of 2-spheres with k components, each of which is separating, then $k < 6t$.*

5.4. 2-Spheres

More generally, Kneser proved that if (M, K) is a triangulated 3-manifold and the number of 3-simplices in K is t, then an independent set of 2-spheres in M has strictly less than $6t + 2 \dim H_2(M; \mathbb{Z}_2)$ components.

We first prove a lemma:

Lemma 5.4.4. *If M contains an independent set of 2-spheres with k components, each of which is separating, then for any triangulation (M, K) of M, (M, K) contains an independent set of k normal 2-spheres.*

The proof of this lemma is very similar to the proof of Theorem 5.2.14. One important difference is that here M is allowed to be reducible. This is why Theorem 5.2.14 describes an isotopy of a surface (that rests on the irreducibility of the 3-manifold in question), whereas here we merely state an existence result. Another difference is that we here use the independence of the 2-spheres in place of incompressibility.

Proof. Let (M, K) be a triangulation of M. Let $S = S_1 \sqcup \cdots \sqcup S_k$ be a set of independent 2-spheres in M chosen so as to minimize $(w(S), m(S))$ among all sets of k independent 2-spheres. We assume that S is in general position with respect to the triangulation.

For each 3-simplex $[s] \in K$ and each face (f) of $[s]$, $S \cap |(f)|$ consists of simple arcs and simple closed curves. Suppose that $S \cap |(f)|$ contains simple closed curves and let c be an innermost such curve. Then c bounds a disk D in $|(f)|$ that meets S only in its boundary. Let S_i be the component of S containing c. Call the component of $M \setminus S_i$ not containing D the "outside" of S_i and the component containing D the "inside" of S_i. Here c cuts S_i into two disks, D' and D''.

Set $S'_i = D \cup D'$ and $S''_i = D \cup D''$. Abusing notation slightly, we also denote small disjoint push-offs of S'_i and S''_i to the inside of S_i by S'_i and S''_i. See Figures 5.21 and 5.22. Set $S' = (S \setminus S_i) \sqcup S'_i$ and $S'' = (S \setminus S_i) \sqcup S''_i$.

Claim. Either S' or S'' is independent.

Suppose that neither S' nor S'' is independent. Then S'_i lies in the frontier of a punctured 3-sphere B' and S''_i lies in the frontier of a punctured 3-sphere B''. Suppose that S_i is contained in, say, B'. Then by the Schönflies Theorem, the 2-sphere S_i cuts the punctured 3-sphere B' into two punctured 3-spheres, B_1, B_2. Note that S'_i lies in the frontier of one of these punctured 3-spheres, say B_1. The other, B_2, does not meet S'_i and hence is a component of $M \setminus S$. (This is where we use the hypothesis that S_i is separating.) But this contradicts the independence of S. Thus B' does not contain S_i. It follows that B' lies on the "inside" of S_i. Likewise, B'' lies on the "inside" of S_i.

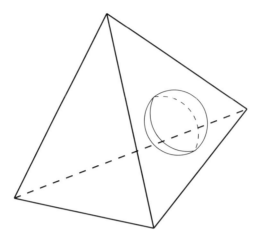

Figure 5.21. *An innermost disk.*

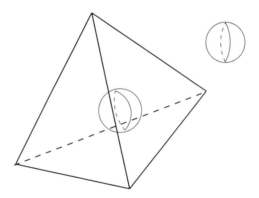

Figure 5.22. *The two resulting spheres.*

It now follows from the construction that B' and B'' meet along D. Hence $B' \cup B'' \cup D$ forms a punctured 3-sphere B in $M \backslash S$. But this also contradicts the fact that S is independent. Thus either S' or S'' is independent.

Suppose S' is independent. Then, after a small isotopy that eliminates $c \cup D$ from $S' \cap |(f)|$, $(w(S'), m(S')) < (w(S), m(S))$. But this contradicts our choice of S so as to minimize $(w(S), m(S))$. Thus for each 3-simplex $[s] \in K$ and each face (f) of $[s]$ the intersection $S \cap |(f)|$ consists of simple arcs. The rest of the proof is identical to the proof of Theorem 5.2.14. □

We now prove Kneser's Theorem. As we will see, the proof relies heavily on understanding how a collection of spheres intersects a given 3-simplex.

5.4. 2-Spheres

Proof of Kneser's Theorem. Suppose M contains a set of k independent 2-spheres, each of which is separating, with $k \geq 6t$. By Lemma 5.4.4, M contains a set $S = S_1 \sqcup \cdots \sqcup S_k$ of k independent normal 2-spheres, each of which is separating.

Let $[s]$ be a 3-simplex in K. A component of $\partial|[s]|\backslash S$ is "good" if it is an annulus that contains no vertex in $|[s]^0|$. At most six components of $\partial|[s]|\backslash S$ are "bad": Indeed, $S \cap |[s]|$ contains at most one type of normal quadrilateral. If, in addition, it contains all types of normal triangles, there will be six "bad" components. Otherwise, there will be fewer "bad" components. See Figure 5.23.

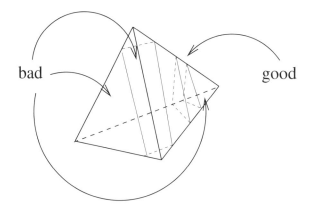

Figure 5.23. *Good components and bad components.*

A component X of $M\backslash S$ is "good" if, for every 3-simplex $[s] \in K$, every component of $X \cap \partial|[s]|$ is good. At most $6t$ components of $M\backslash S$ are "bad". Since $k \geq 6t$, $M\backslash S$ has at least $6t+1$ components. Thus there is at least one "good" component. A "good" component, X, is the union of regions homeomorphic to (triangle)$\times I$ and (quadrilateral)$\times I$. It is not hard to see that then $X = \mathbb{S}^2 \times I$. Specifically, the (triangle)$\times I$ and (quadrilateral)$\times I$ regions constitute an I-bundle over a surface B and this I-bundle is bounded by 2-spheres. Thus B is covered by the 2-sphere and must hence be either a projective plane (in the case of a twisted I-bundle) or a 2-sphere (in the case of a product).

We leave it as an exercise to show that the twisted I-bundle over $\mathbb{R}P^2$ is a punctured $\mathbb{R}P^3$. It follows that $X = \mathbb{S}^2 \times I$, but this contradicts the fact that the set S is independent. Hence $k < 6t$. □

To recover the stronger form of Kneser's Theorem, we need to consider the case where X is a twisted I-bundle over the projective plane. In this case X is bounded by a 2-sphere. Hence it is a summand of M. Moreover, its homology is a direct summand of the homology of M and contributes to $\dim H_2(M; \mathbb{Z}_2)$.

An analogous argument also proves the following theorem:

Theorem 5.4.5 (Haken's Theorem). *Let (M, K) be a triangulated 3-manifold. Suppose that the number of 3-simplices in K is t. If M contains a set of surfaces $F = F_1 \sqcup \cdots \sqcup F_k$ such that no component of $M \backslash F$ is homeomorphic to (surface)$\times I$, then $k \leq 6t + 2 \dim H_2(M; \mathbb{Z}_2)$.*

Actually, Haken gave a bound involving $61t$. Many arguments for this theorem have been given and the bound has improved over time. Combined, the above two theorems are known as Kneser-Haken finiteness.

Exercises

Exercise 1. Show that the twisted I-bundle over $\mathbb{R}P^2$ is a punctured $\mathbb{R}P^3$.

Exercise 2. How many non-isotopic incompressible surfaces are contained in the 3-torus? How large can a collection of disjoint non-isotopic incompressible surfaces in the 3-torus be?

Exercise 3. Generalize Haken finiteness to the case of incompressible surfaces with boundary properly embedded in 3-manifolds with boundary. Caution: There is a hypothesis that must be added to obtain a generalization.

5.5. Prime Decompositions

We will show that every orientable 3-manifold has a prime decomposition that is unique up to a reordering of its factors. (A prime 3-manifold is considered to have a prime decomposition with just itself as a factor.) Beware that there are subtleties concerning the "uniqueness" of this prime decomposition. The prime decomposition is "unique" in the sense that the factors are unique up to reordering. The decomposing spheres for the factorization are far from unique (even up to isotopy). The existence of a prime decomposition can be established using Kneser's Theorem:

Theorem 5.5.1. *Every compact 3-manifold can be expressed as a connected sum of a finite number of prime factors.*

Proof. Let M be a compact 3-manifold and let (M, K) be a triangulation of M. Let S be an independent set of k normal 2-spheres, each of which is separating, chosen so that k is maximal among all such sets. This is possible by Kneser's Theorem. Then

$$M \backslash S = \hat{M}_1 \sqcup \cdots \sqcup \hat{M}_{k+1}.$$

5.5. Prime Decompositions

Denote the compact 3-manifold obtained by attaching 3-balls along the frontier of $\hat{M}_i$ by M_i. We need to show that M_i is prime. Indeed, suppose to the contrary that there is a separating essential 2-sphere $\tilde{S}$ in M_i. Then $S \sqcup \tilde{S}$ is a set of $k+1$ normal 2-spheres, each of which is essential and separating. Since S was chosen to have a maximal number of components among independent sets of 2-spheres, each of which is separating, $S \sqcup \tilde{S}$ is not independent. Thus a component of $\hat{M}_i \backslash \tilde{S}$ is a punctured 3-sphere. But then the corresponding component of $M_i \backslash \tilde{S}$ is a 3-ball. Since $\tilde{S}$ is essential, this is a contradiction. Hence M_i is prime. Thus

$$M = M_1 \# \cdots \# M_{k+1}$$

is as required. $\square$

Definition 5.5.2. A component of an independent set of separating 2-spheres is called a *decomposing sphere*.

Recall the discussion of the operation of connected sum in Chapter 1. In considering prime decompositions of 3-manifolds, we must take into account that there can be several independent sets of 2-spheres providing a prime decomposition of a given 3-manifold. I.e., collections of decomposing spheres are not unique. Nevertheless, prime decompositions of 3-manifolds are unique in the sense expressed in the following theorem:

Theorem 5.5.3 (Uniqueness of prime decompositions). *Let M be a compact orientable 3-manifold. If $M = M_1 \# \cdots \# M_k = N_1 \# \cdots \# N_l$ are prime decompositions, then $k = l$ and, after reordering, M_i is homeomorphic to N_i, for $i = 1, \ldots, l$.*

Before proving the uniqueness of prime decomposition, we prove a lemma:

Lemma 5.5.4. *Let Σ be a non-separating 2-sphere and let S be a separating 2-sphere in the 3-manifold M. Let c be an innermost component of $\Sigma \cap S$ in S. Let D be the component of $S \backslash c$ that is disjoint from Σ and let D', D'' be the components of $\Sigma \backslash c$. Set $\Sigma' = D \cup D' \cup c$ and $\Sigma'' = D \cup D'' \cup c$. Then either Σ' or Σ'' is non-separating.*

Proof. Recall Remark 3.3.3. Since Σ is non-separating, there is a simple closed curve α in M that meets Σ exactly once and is transverse to $\Sigma \cup S$. The number of points in $\alpha \cap (\Sigma' \cup \Sigma'')$ is equal to the sum of the number of points in $\alpha \cap \Sigma' = \alpha \cap (D \cup D')$ and $\alpha \cap \Sigma'' = \alpha \cap (D \cup D'')$. Modulo 2, this sum is equal to the sum of the number of points in $\alpha \cap D'$ and $\alpha \cap D''$ which is equal to $\alpha \cap (D' \cup D'')$ and hence equal to the number of points in $\alpha \cap \Sigma$. Thus there are an odd number of points in either $\alpha \cap \Sigma'$ or $\alpha \cap \Sigma''$, say in $\alpha \cap \Sigma'$. The existence of the simple closed curve α that meets Σ' in an odd number of points shows that Σ' is non-separating. $\square$

Proof of Theorem 5.5.3.

Case 1. M contains no non-separating 2-spheres.

In this case, each M_i and each N_j is irreducible. Let $S = S_1 \sqcup \cdots \sqcup S_{k-1}$ be a set of 2-spheres such that $M \backslash S = M_1^* \sqcup \cdots \sqcup M_k^*$, where M_i^* is a punctured copy of M_i for $i = 1, \ldots, k$, and let Σ be a 2-sphere such that $M \backslash \Sigma = N_1^* \sqcup M'$, for N_1^* a once-punctured N_1.

We may assume that Σ and S have been chosen so that the number of components of $\Sigma \cap S$ is minimal over all spheres isotopic to Σ and all sets of 2-spheres that decompose M into 3-manifolds homeomorphic to $M_1, \ldots, M_k$. Suppose that $\Sigma \cap S \neq \emptyset$. Let c be an innermost component of $\Sigma \cap S$ in Σ. Let D be the component of $\Sigma \backslash c$ that is disjoint from S. In particular, $D \subset M_j^*$ for some j. Let S_i be the component of S containing c.

Note that $S_i \backslash c$ consists of two disks, D', D''. Set $S' = D \cup D'$, $S'' = D \cup D''$. Recall that M_j is irreducible. Thus S', S'' bound 3-balls B', B'' in M_j. If the interiors of B', B'' are disjoint, then $S_i = \partial(B' \cup B'')$, but this is impossible, since S_i is a decomposing sphere. Thus, since M_j is not $\mathbb{S}^3$, it must be the case that either $B' \subset B''$ or $B'' \subset B'$, say $B' \subset B''$. Now we construct a new set of 2-spheres S^* by setting $S_i^* = S''$ and $S_j^* = S_j$ for $j \neq i$. Then $S^* = S_1^* \sqcup \cdots \sqcup S_{k-1}^*$ also has the property that $M \backslash S^*$ is the disjoint union of punctured copies of $M_1, \ldots, M_k$ but the number of components of $\Sigma \cap S^*$ is strictly less than the number of components of $\Sigma \cap S$. This contradicts our choice of Σ and S. Thus $\Sigma \cap S = \emptyset$.

Suppose now that a component of S, say S_c, lies in N_1^*. Since N_1 is irreducible, S_c bounds a 3-ball in N_1. On the other hand, since S_c is essential in M, it does not bound a 3-ball in M and hence does not bound a 3-ball in N_1^*. It follows that S_c bounds a punctured 3-ball in N_1^* with frontier a subset of $\partial N_1^* \cap \Sigma$. A component, M_c^*, of $M \backslash S$ adjacent to S_c then contains a punctured copy of N_1. Likewise, if no component of S lies in N_1^*, then $N_1^* \subset M_c^*$ for some c. It follows that for $j = 1, \ldots, l$, a punctured copy of N_j lies in M_{c_j} for some c_j. Furthermore, since M_{c_j} is prime, M_{c_j} does not contain punctured copies of other factors. By a symmetric argument, for $i = 1, \ldots, k$, a punctured copy of M_i lies in N_{p_i} for some p_i and N_{p_i} does not contain punctured copies of other factors. Thus $k = l$ and, after reordering, $M_i = N_i$.

Case 2. M contains a non-separating 2-sphere $\tilde{S}$.

Let S be as above. We may assume that $\tilde{S}$ is chosen so that the number of components of $\tilde{S} \cap S$ is minimal over all separating 2-spheres in M. Suppose that $\tilde{S} \cap S \neq \emptyset$. Let c be an innermost component of $\tilde{S} \cap S$ in S. Proceeding as in Lemma 5.5.4, we obtain $\tilde{S}', \tilde{S}''$ such that $\tilde{S}'$ or $\tilde{S}''$, say $\tilde{S}'$, is non-separating. However, the number of components of $\tilde{S}' \cap S$ is strictly less than the number of components of $\tilde{S} \cap S$, contradicting the assumed minimality. Thus $\tilde{S} \cap S = \emptyset$. Let M_i^* be a component of $M \backslash S$ containing the non-separating 2-sphere $\tilde{S}$. By Theorem 3.3.4, $M_i = \mathbb{S}^2 \times \mathbb{S}^1$. Similarly, we show that for some j, $N_j = \mathbb{S}^2 \times \mathbb{S}^1$. Thus after reordering, $M_i = N_i$. Splitting off summands equal to $\mathbb{S}^2 \times \mathbb{S}^1$ reduces the argument to the case where M contains no non-separating 2-sphere. $\square$

For non-orientable 3-manifolds prime factorizations are not unique. More specifically, if M is non-orientable and $M = M_1 \#(\mathbb{S}^2 \times \mathbb{S}^1)$, then it is also the case that $M = M_1 \#(\mathbb{S}^2 \tilde{\times} \mathbb{S}^1)$ and vice versa. However, this prime factorization becomes unique if we decree that reducible prime summands for non-orientable 3-manifolds always be $\mathbb{S}^2 \tilde{\times} \mathbb{S}^1$. See [**63**].

Recall that a 3-manifold is non-orientable if and only if it contains a submanifold homeomorphic to (Möbius band)$\times I$. Twisted I-bundles are interesting when considering the question of orientation. The Möbius band can be thought of as a twisted I-bundle over the circle. In particular, the circle is orientable but the Möbius band is not. The projective plane is not orientable, but the twisted I-bundle over the projective plane is orientable. Its boundary is a 2-sphere.

Exercises

Exercise 1. Which prime 3-manifold contains the twisted I-bundle over the projective plane?

Exercise 2. How many different decomposing spheres does $(\mathbb{S}^2 \times \mathbb{S}^1) \# \mathbb{T}^3$ contain (up to isotopy)?

5.6. Recognition Algorithms

Corollary 5.3.7 establishes the existence of an algorithm to determine whether an irreducible 3-manifold contains an incompressible surface. In [**52**], Haken discusses how to use normal surface theory to decide whether or not two Haken 3-manifolds are homeomorphic. See also [**156**], [**61**], or [**96**]. Since the 3-sphere is not a Haken manifold, that discussion does not address the question as to whether it is possible to recognize the 3-sphere. It is in fact possible to recognize the 3-sphere, but this was proved much later by Rubinstein. Rubinstein sketched an argument that was completed by Thompson; see [**152**]. Matveev gave an alternative treatment; see [**96**]. The algorithm relies on an extension of normal surface theory called "almost" normal surface theory. A recognition algorithm for all closed orientable 3-manifolds was given by Sela (see [**147**]) and Manning (see [**92**]).

Theorem 5.6.1. *There is an algorithm to determine whether or not a closed 3-manifold is $\mathbb{S}^3$.*

Definition 5.6.2. A disk in a 3-simplex that is bounded by a normal curve of length 8 is called an *almost normal disk*. A surface in a triangulated 3-manifold is *almost normal* if it is normal away from exactly one almost normal disk.

Remark 5.6.3. In some contexts, almost normal surfaces are defined in a broader sense. There, they meet exactly one 3-simplex in normal disks and either one almost normal disk or two normal disks that are joined by a tube. We will not use this more general definition here.

Remark 5.6.4. Note that the set of almost normal surfaces is disjoint from the set of normal surfaces!

A crucial step in the proof of Theorem 5.6.1 establishes that a closed triangulated 3-manifold that contains an almost normal 2-sphere and no normal 2-spheres is necessarily $\mathbb{S}^3$. As it turns out, if an irreducible triangulated 3-manifold (M, K) contains a normal 2-sphere, then M admits a triangulation (M, K') with fewer 3-simplices. Conversely, a triangulation of $\mathbb{S}^3$ contains either an almost normal 2-sphere or a normal 2-sphere. Whether or not a 3-manifold contains an almost normal 2-sphere can be determined via an algorithm similar to Haken's algorithm.

The following theorem and corollary rely heavily on Theorem 5.6.1.

Theorem 5.6.5. *If the closed triangulated 3-manifold (M, K) contains an incompressible 2-sphere or projective plane, then it contains an incompressible fundamental 2-sphere or projective plane.*

Proof. Let F be the 2-sphere or projective plane. By isotoping F as in Theorem 5.2.14 we arrange for F to be normal. We may assume that F has been chosen so that $w(F)$ is minimal among all incompressible normal 2-spheres and projective planes in M and that F is in general position with respect to K.

Suppose F is not fundamental. By Lemma 5.3.5 there are connected normal surfaces G, H such that $F = G + H$, $w(G) > 0$, $w(H) > 0$, $w(F) = w(G) + w(H)$, and $\chi(F) = \chi(G) + \chi(H)$. Moreover, by Theorem 5.3.6, G and H are incompressible.

Case 1. $\chi(F) = 2$; i.e., F is a 2-sphere.

Up to renaming of G, H, there are only two possibilities: (1) $\chi(G) = 2$ and $\chi(H) = 0$ or (2) $\chi(G) = 1$ and $\chi(H) = 1$. In both of these cases, G is either a 2-sphere or projective plane and $w(G) < w(F)$. This contradicts the minimality of $w(F)$.

Case 2. $\chi(F) = 1$; i.e., F is a projective plane.

Up to renaming of G, H, there are only two possibilities: (1) $\chi(G) = 2$ and $\chi(H) = -1$ or (2) $\chi(G) = 1$ and $\chi(H) = 0$. Thus G is either a 2-sphere or projective plane and $w(G) < w(F)$, contradicting the assumed minimality of $w(F)$.

Thus F is not the sum of other incompressible surfaces under the + operation. It follows that F is fundamental. In particular, M contains a fundamental 2-sphere or projective plane. $\square$

Corollary 5.6.6. *There is an algorithm to decide whether or not a closed 3-manifold contains an incompressible 2-sphere or projective plane.*

The idea for the proof of this corollary is the following: Solve the Diophantine system corresponding to the 3-manifold as in Theorem 5.3.2. Recall that this gives a finite set of solutions that generates all solutions of the Diophantine system. Moreover, this finite set of solutions contains the set of fundamental solutions. These can then be found by discarding those elements of this finite set that are linear combinations with non-negative integral coefficients of other elements. Now for each fundamental solution, consider a representative of the corresponding normal isotopy class of fundamental surfaces and compute its Euler characteristic. If the Euler characteristic of a fundamental surface is 1, then the fundamental surface is a projective plane. If it is 2, then it is a 2-sphere. In each such case, check whether or not the projective plane or 2-sphere is incompressible. In the case of the 2-sphere, this latter step requires the 3-sphere recognition algorithm, Theorem 5.6.1.

In a similar vein, Haken proved the following theorem:

Theorem 5.6.7 (Haken). *If the triangulated 3-manifold (with boundary) M contains an essential disk with specified boundary and if it contains no projective planes, then it contains a fundamental disk with boundary isotopic to the boundary specified.*

The system of Diophantine equations employed in the proof of this theorem must incorporate the information concerning the specified boundary. Specifically, if we are concerned with disks with specified boundary c, we find a triangulation for M such that each 2-dimensional face of a 3-simplex lying in the boundary of M meets c in at most one normal arc. This can easily be accomplished by choosing any given triangulation of M, isotoping c, and taking barycentric subdivisions of the triangulation if necessary. Everything else remains the same. Note that the fundamental solutions of this "relative" system of Diophantine equations will correspond to fundamental surfaces that are either closed surfaces or have boundary components that lie in the boundary specified.

The application Haken had in mind is the following:

Corollary 5.6.8. *There is an algorithm to decide whether or not a knot $K \subset \mathbb{S}^3$ is the unknot.*

Proof. Let $\eta(K)$ be an open regular neighborhood of K and set $C(K) = \mathbb{S}^3 \backslash \eta(K)$. By Exercise 1 of Section 4.3, we know that a knot is the unknot if and only if $C(K)$ contains an essential disk with longitudinal boundary. Since the complement of a knot in $\mathbb{S}^3$ contains no projective planes, Haken's Theorem implies that containing an essential disk with longitudinal boundary is equivalent to containing a fundamental disk with longitudinal boundary. (Since a disk with longitudinal boundary is non-separating, it is necessarily essential.) The existence of a fundamental disk with longitudinal boundary can be established or ruled out by triangulating the knot complement appropriately and finding the fundamental solutions of the corresponding Diophantine system. □

In [**85**] and [**84**], Li used the theory of almost normal surfaces and other techniques to prove a conjecture of Waldhausen: An irreducible atoroidal 3-manifold admits only finitely many distinct Heegaard splittings of a given genus.

Exercises

Exercise 1. Show that if $F = G + H$ and H is an inessential sphere, then F is isotopic to G.

Exercise 2. Devise an algorithm to generate all Seifert surfaces of a knot complement.

Exercise 3. Devise an algorithm to calculate the genus of a knot K (that is, the smallest possible genus of a Seifert surface for K).

5.7. PL Minimal Surfaces**

Triangulations can be used to endow manifolds with metric structures similar to those constructed in the context of differentiable manifolds. For instance, the sophisticated analytic theory of minimal surfaces has an analog in the context of triangulated 3-manifolds. All that is needed here is some understanding of metric structures on surfaces and in fact only on triangles. Because this provides a more accessible version of minimal surface theory, we include a brief overview here. PL minimal surface theory was developed by Jaco and Rubinstein. For more details, see [**70**], [**69**], and [**38**].

5.7. PL Minimal Surfaces

Given a compact triangulated 3-manifold (M, K), the 2-simplices can be endowed with a metric g of negative curvature with geodesic edges and with vertices at infinity. Moreover, this can be accomplished in such a way that the metrics on these simplices match up along 1-simplices where they meet and the resulting path-metric on K^2 is complete. Call such a triple (M, K, g) a *3-manifold with metric triangulation*. Jaco and Rubinstein employ hyperbolic metrics on the 2-simplices of the triangulation but point out that this is not necessary.

Recall that the *weight*, denoted by $w(F)$, of a surface F in (M, K, g) is the number of points in $F \cap K^1$. The *measure* of F, denoted by $l(F)$ is the total length $l(F \cap K^2)$, that is, the sum of the lengths of all the arcs in which F meets the 2-simplices of the triangulation. (Do not confuse length with measure.) The surface F is *PL-minimal* if it minimizes l for small isotopies of F. It is *PL least area* if it minimizes (w, l) in its isotopy class (according to the dictionary order).

We list several theorems about PL least area surfaces. All are variations of theorems in [**69**]. We have formulated them in terms of isotopy classes of embedded surfaces, though the theory has traditionally been formulated in terms of homotopy classes of immersed surfaces. (The arguments carry over verbatim to the context of isotopy classes of embedded surfaces.)

The following existence and uniqueness result provides a canonical choice of representative for an isotopy class of surfaces. This type of canonical representative proves useful in many arguments in 3-dimensional topology. For a short proof of the uniqueness part of this theorem, see [**113**].

Theorem 5.7.1. *Let E be a compact orientable irreducible 3-manifold endowed with a metric triangulation and let F be a compact incompressible surface in E. Then there exists a unique PL least area representative in the isotopy class of F.*

The theorem below tells us that lifts of PL least area surfaces from a 3-manifold to a covering space are PL least area and also that projections of PL least area surfaces from the covering space to the 3-manifold are PL least area, provided that the projection is injective on the surface.

Theorem 5.7.2. *Let E be a compact orientable irreducible 3-manifold endowed with a metric triangulation and let M be any covering space of E. Let F_E be a 2-sided compact incompressible surface in E whose lift F_M to M is also compact. Then F_E is PL least area if and only if F_M is PL least area.*

Minimality arguments are used frequently in 3-dimensional topology, most notably arguments relying on surfaces that have been isotoped to intersect in a minimal number of components. The two results below establish the fact that PL least area surfaces necessarily intersect in the smallest number of components possible in their isotopy classes.

Theorem 5.7.3. *Let E be a compact orientable irreducible 3-manifold endowed with a metric triangulation and let F_1 and F_2 be PL least area incompressible surfaces in E with disjoint representatives in their isotopy classes. Then either F_1 and F_2 are disjoint or they coincide.*

Theorem 5.7.4. *If F_1, F_2 are PL least area incompressible surfaces in the compact orientable irreducible 3-manifold E endowed with a metric triangulation, then their number of components of intersection is minimal over all representatives of the isotopy classes of the surfaces.*

Exercises

Exercise 1*. Use PL minimal surface theory to prove that a covering space of a 3-manifold M is irreducible if and only if M is irreducible.

Exercise 2. Prove that the double curve sum of two minimal genus Seifert surfaces contains two minimal genus Seifert surfaces.

Exercise 3. Suppose that the knot K has only finitely many isotopy classes of minimal genus Seifert surfaces. Prove that a PL least area minimal genus Seifert surface for K whose area is strictly less than that of all other Seifert surfaces for K is disjoint from at least one non-isotopic Seifert surface of K.

Chapter 6

Heegaard Splittings

A Heegaard splitting is, roughly speaking, a splitting of a 3-manifold into two simple pieces called handlebodies. Below, we will state a theorem of Bing that establishes that every 3-manifold admits such a splitting. It turns out, however, that "the whole is more than the sum of its parts". Specifically, though the two handlebodies are well understood as 3-manifolds, their relative positioning remains a worthy object of study. For the origins of the subject, see [60]. For an overview, see [133].

In this chapter, we will be interested in Heegaard splittings and generalized Heegaard splittings. We will see basic theorems concerning Heegaard splittings, structural and classification theorems for Heegaard splittings, and generalized Heegaard splittings and applications of such theorems.

6.1. Handle Decompositions

A handle decomposition of a 3-manifold is a particular way to build the 3-manifold. Handle decompositions exist for manifolds of any dimension. For smooth manifolds, this is a consequence of the fact that every manifold admits a Morse function. For more information, see Appendix B.

Definition 6.1.1. In the 3-dimensional setting, a *k-handle* is a 3-ball, thought of as $[0,1]^3$, that is attached (to some preexisting submanifold) along $[0,1]^{3-k} \times \partial[0,1]^k$. (More generally, in the n-dimensional setting, a *k-handle* is an n-ball, thought of as $[0,1]^n$, that is attached (to some preexisting submanifold) along $[0,1]^{n-k} \times \partial[0,1]^k$.) See Figure 6.1.

Concretely, a 0-handle is a 3-ball attached to the empty set. We think of this as a 3-ball appearing out of nowhere. A 1-handle is a 3-ball attached along $[0,1]^2 \times \partial[0,1]$. We think of this as a solid cylinder attached along its two bounding disks. A 2-handle is a 3-ball attached along $[0,1] \times \partial[0,1]^2$. We think of this as a solid cylinder attached along the annular portion of its boundary. Finally, a 3-handle is a 3-ball attached along $\partial[0,1]^3$, i.e., along its entire boundary. We think of this as a 3-ball that is being used to fill an existing hole.

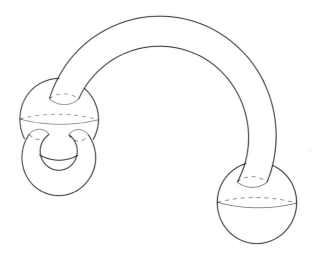

Figure 6.1. *Two 0-handles and two 1-handles.*

Definition 6.1.2. A *handle decomposition* of a 3-manifold M is a sequence of 0-handles, 1-handles, 2-handles, and 3-handles whose union is M.

In Appendix B we state a theorem that guarantees the existence of Morse functions on a manifold. We also explain how this guarantees the existence of handle decompositions of the manifold. In addition, it follows from technical results on Morse functions that manifolds have handle decompositions in which all 0-handles are attached before all 1-handles, all 1-handles are attached before all 2-handles, all 2-handles are attached before all 3-handles, and so on. We will sometimes, but not always, have such handle decompositions in mind. The subtle issue of which handle decompositions describe the same manifold is treated in [**77**].

Definition 6.1.3. The *core* of a 3-dimensional k-handle is $\{\frac{1}{2}\}^{3-k} \times [0,1]^k$. The *cocore* is $[0,1]^{3-k} \times \{\frac{1}{2}\}^k$. See Figures 6.2 and 6.3.

Remark 6.1.4. In the context of 3-manifolds, a k-handle is dual to a $(3-k)$-handle in the following sense: A k-handle in a closed 3-manifold is a 3-ball attached to some preexisting submanifold along a portion of its boundary.

6.1. Handle Decompositions

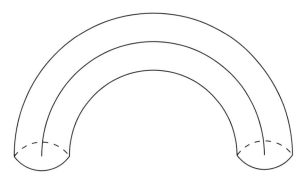

Figure 6.2. *The core of a 3-dimensional 1-handle.*

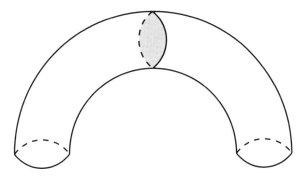

Figure 6.3. *The cocore of a 3-dimensional 1-handle.*

The remaining portion of the 3-manifold will be attached to the boundary of the resulting 3-manifold.

Given a handle decomposition of the 3-manifold M (thought of as a procedure for building M), we can "reverse the procedure" to obtain the dual handle decomposition: We attach what were formerly 3-handles first, but they are now attached to nothing and hence are 0-handles. We then attach what were formerly 2-handles, but these are now attached along two disks and hence are 1-handles, etc. The core of a k-handle in the original handle decomposition is the cocore of the corresponding $(3-k)$-handle when we reverse the procedure.

Definition 6.1.5. A *handlebody* is a compact connected orientable 3-manifold with boundary that possesses a handle decomposition consisting of 0-handles and 1-handles. The *genus* of a handlebody is the genus of its boundary.

A collection of *meridian disks* for a handlebody is a collection of disks that cut the handlebody into 3-balls.

Example 6.1.6. The handlebody pictured in Figure 6.1 is a genus 1 handlebody. We also refer to this handlebody as a solid torus.

Every handlebody has a collection of meridian disks: Start with a handle decomposition consisting of only 0-handles and 1-handles and take the set of cocores of the 1-handles in this handle decomposition. See Figure 6.4.

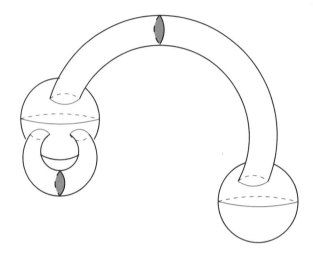

Figure 6.4. *A set of meridian disks for a handlebody.*

Definition 6.1.7. A *Heegaard splitting* of a closed 3-manifold M is a decomposition $M = V \cup_S W$ such that

- V, W are handlebodies and
- $S = \partial V = \partial W$.

Here S is called the *splitting surface* of $M = V \cup_S W$.

Two Heegaard splittings are considered *equivalent* if their splitting surfaces are isotopic. The *genus* of a Heegaard splitting is the genus of S.

We think of the handlebodies V, W being *glued* along S to create M.

Example 6.1.8. The 3-sphere has a genus 0 Heegaard splitting: We think of $\mathbb{S}^3$ as consisting of those points in $\mathbb{R}^4$ that have distance 1 from the origin. The subspace given by $w = 0$ intersects $\mathbb{S}^3$ in a 2-sphere. This 2-sphere separates $\mathbb{S}^3$ into two 3-balls, each a handlebody of genus 0.

Example 6.1.9. If two solid tori T_1 and T_2 are glued together in such a way that a meridian of T_1 is identified to a longitude of T_2, then the result is $\mathbb{S}^3$. This describes a genus 1 Heegaard splitting of $\mathbb{S}^3$.

6.1. Handle Decompositions

Example 6.1.10. As above, consider two solid tori T_1, T_2. If the two solid tori are glued together in such a way that a meridian of T_1 is identified to a (p,q)-torus knot of T_2, then the result is a lens space, more specifically, a genus 1 Heegaard splitting of the lens space $L(p,q)$.

Example 6.1.11. We can describe a Heegaard splitting of the 3-torus $\mathbb{S}^1 \times \mathbb{S}^1 \times \mathbb{S}^1$ as follows: We think of the 3-torus as a quotient space obtained by identifying opposite sides of a cube. Denote a regular neighborhood of the 1-skeleton of the cube by V and denote the closure of its complement by W. Then both V and W are handlebodies. The two handlebodies meet in the surface S of genus 3. See Figure 6.5.

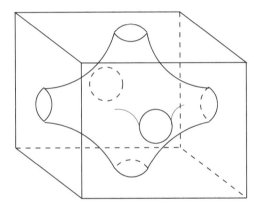

Figure 6.5. *A Heegaard splitting of the 3-torus.*

The following theorem is due to E. Moise; see [**101**].

Theorem 6.1.12 (Moise)**.** *Every closed orientable 3-manifold admits a Heegaard splitting.*

We offer two proofs of this theorem:

Proof (Version 1). This theorem follows from two facts: (1) Every 3-manifold admits a handle decomposition in which all 0-handles are attached before all 1-handles, which in turn are attached before all 2-handles, which in turn are attached before all 3-handles; (2) 2-handles are dual to 1-handles and 3-handles are dual to 0-handles.

The 0-handles and 1-handles in a handle decomposition of this type provide one handlebody, and the 2-handles and 3-handles, dually, provide the other. □

Proof (Version 2). Let M be a closed 3-manifold. Then M admits a triangulation (M, K). Set $V = N(K^1)$. Then V is a handlebody. It is not too hard to see that the closure, W, of the complement of V is also a handlebody. See Figure 6.6. □

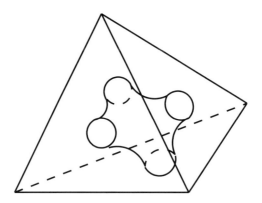

Figure 6.6. *Local picture of Heegaard splitting.*

For a broader view of Heegaard splittings and how they relate to other structures on 3-manifolds, see [**83**] and [**90**].

Exercises

Exercise 1. Prove that $\mathbb{S}^3$ is the only 3-manifold with a genus 0 Heegaard splitting.

Exercise 2. Prove that all handlebodies of genus g are homeomorphic.

Exercise 3. Prove that, for $g > 0$, there are handlebodies of a genus g in $\mathbb{S}^3$ that are not isotopic.

Exercise 4. Prove that the construction in Version 2 of the proof of Theorem 6.1.12 describes a Heegaard splitting of the 3-manifold.

6.2. Heegaard Diagrams

You showed in the exercises for the previous section that all handlebodies of a given genus are homeomorphic. It follows that, given a Heegaard splitting $M = V \cup_S W$ of genus g, we can visualize the genus g handlebody V in $\mathbb{R}^3$. To reconstruct M together with its Heegaard splitting we need only know how to attach W to the outside of V. This reduces to understanding how a collection of meridian disks for W is attached to ∂V. (By Alexander's Theorem, there is then only one way to attach the 3-handles $W\backslash$(meridian disks).)

Note that the embedding of V in $\mathbb{R}^3$ specifies a collection of meridian disks for V (though not uniquely). These, in turn, specify (and are specified

6.2. Heegaard Diagrams

by), a collection of curves on ∂V. This gives us a framework in which to represent the curves along which the collection of meridian disks for W is attached along ∂V.

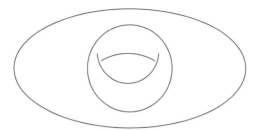

Figure 6.7. *A Heegaard diagram for the 3-sphere.*

Definition 6.2.1. A *Heegaard diagram* is a closed orientable surface S of genus g that is equipped with two collections of essential simple closed curves $\{v_1, \ldots, v_g\}$ and $\{w_1, \ldots, w_g\}$.

Remark 6.2.2. We think of obtaining a 3-manifold M by attaching 2-handles to the inside of $S \times I$ along $\{v_1, \ldots, v_g\}$ and to the outside of $S \times I$ along $\{w_1, \ldots, w_g\}$. Attaching 2-handles to the inside of $S \times I$ along $\{v_1, \ldots, v_g\}$ yields a handlebody V. An embedding of V in $\mathbb{R}^3$ specifies a collection of curves $\{v'_1, \ldots, v'_g\}$, though not uniquely. However, attaching 2-handles to the inside of $\partial V \times I = S \times I$ along any such collection of curves and 2-handles to the outside of $S \times I$ along curves $\{w_1, \ldots, w_g\}$ yields the same 3-manifold. For this reason we only indicate one collection of curves, $\{w_1, \ldots, w_g\}$, in Figures 6.7 and 6.8.

Example 6.2.3. Figure 6.7 gives a Heegaard diagram of $\mathbb{S}^3$.

Example 6.2.4. Figure 6.8 gives a Heegaard diagram of the lens space $L(2,1)$.

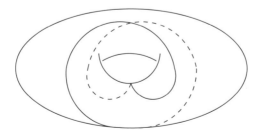

Figure 6.8. *A Heegaard diagram for real projective 3-space.*

A more traditional way to exhibit a Heegaard diagram is to obtain a punctured surface by cutting along the curves $v_1, \ldots, v_g$ and presenting this punctured surface along with the remnants of $w_1, \ldots, w_g$. The boundary components of the resulting punctured surface are then numbered so as to allow a reconstruction of the genus g surface.

Example 6.2.5. Figure 6.9 gives a traditional Heegaard diagram of the 3-torus.

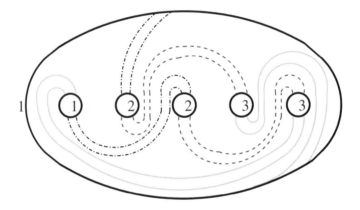

Figure 6.9. *A traditional Heegaard diagram for the 3-torus.*

Exercises

Exercise 1. Draw a Heegaard diagram representing $\mathbb{S}^3$.

Exercise 2. Draw a Heegaard diagram representing a manifold obtained by Dehn surgery on the figure 8 knot.

Exercise 3. Prove that $\mathbb{R}P^3$ is homeomorphic to $L(2,1)$.

6.3. Reducibility and Stabilization

Given two 3-manifolds with Heegaard splittings, we can find a Heegaard splitting of the connected sum of the two 3-manifolds. We do this by considering the pairwise connected sum of the 3-manifolds relative to their splitting surfaces. Furthermore, given one 3-manifold with a Heegaard splitting, we can find other Heegaard splittings for this manifold. This is accomplished, for instance, by the process, described below, known as stabilization.

Definition 6.3.1. Let $M = V \cup_S W$ be a Heegaard splitting and let $\mathbb{S}^3 = V' \cup_T W'$ be the standard genus 1 Heegaard splitting of $\mathbb{S}^3$. The pairwise connected sum $(M, S) \# (\mathbb{S}^3, T)$ defines a Heegaard splitting $M = \tilde{V} \cup_{\tilde{S}} \tilde{W}$ called an *elementary stabilization* of $M = V \cup_S W$. A Heegaard splitting is called a *stabilization* of $M = V \cup_S W$ if it is obtained from $M = V \cup_S W$ by performing a finite number of elementary stabilizations.

6.3. Reducibility and Stabilization

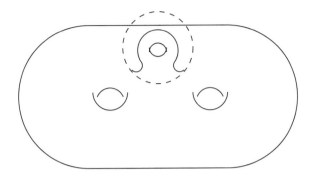

Figure 6.10. *A stabilization.*

Theorem 6.3.2 (Reidemeister-Singer). *Any two Heegaard splittings of a 3-manifold M become equivalent after a finite number of stabilizations.*

This theorem can be proved by considering Morse functions. In Appendix B, we discuss how Heegaard splittings correspond to Morse functions. As it turns out, any two Morse functions are related by a sequence of two moves, namely stabilization and exchanging levels of critical points. The latter can be understood via handle decompositions: The order in which two handles in the handle decomposition are attached is interchanged. Thus this latter move changes the handle decomposition corresponding to the Morse function but not the corresponding Heegaard splitting. Morse functions were studied extensively by J. Cerf in his analysis of smooth real-valued functions on smooth manifolds in [**26**].

We will consider two important properties for Heegaard splittings. These properties will allow us to establish two theorems. One consequence will be that a Heegaard splitting of a connected sum of 3-manifolds factors into Heegaard splittings of the summands.

Definition 6.3.3. A Heegaard splitting $M = V \cup_S W$ is *reducible* if there is an essential simple closed curve $c \subset S$ and disks $D \subset V$, $E \subset W$ with $\partial D = \partial E = c$. Alternatively, $M = V \cup_S W$ is *reducible* if there is a 2-sphere Σ in M such that $\Sigma \cap S$ is an essential simple closed curve. A Heegaard splitting is *irreducible* if it is not reducible.

A Heegaard splitting $M = V \cup_S W$ is *weakly reducible* if there are disks $D \subset V$, $E \subset W$ such that ∂D, ∂E are essential in S and $\partial D \cap \partial E = \emptyset$. A Heegaard splitting is *strongly irreducible* if it is not weakly reducible.

Note that a reducible Heegaard splitting is weakly reducible since ∂D can be isotoped to be disjoint from ∂E. It follows that a strongly irreducible Heegaard splitting is irreducible. Strong irreducibility has proved to be a useful concept in the study of Heegaard splittings. In many ways, the splitting surface of a strongly irreducible Heegaard splitting behaves like an incompressible surface.

Let us briefly consider the Heegaard splittings we encountered in the previous section. The Heegaard splitting of $\mathbb{S}^3$ of genus 0 is irreducible. The Heegaard splitting of $\mathbb{S}^3$ of genus 1 is also irreducible. The Heegaard splittings described for lens spaces are irreducible unless the 3-manifold in question turns out to be $\mathbb{S}^2 \times \mathbb{S}^1$. In the latter case the Heegaard splitting is reducible. The genus 3 Heegaard splitting of the 3-torus is irreducible but weakly reducible. See Figure 6.11.

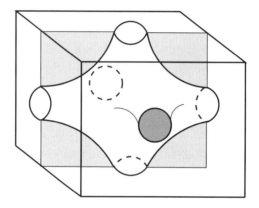

Figure 6.11. *The genus 3 Heegaard splitting of the 3-torus is weakly reducible.*

Interestingly, the Heegaard splitting of genus 1 of $\mathbb{S}^3$ is stabilized, though it is irreducible. It is the unique example of a stabilized Heegaard splitting that is irreducible. The standard Heegaard splitting of $\mathbb{S}^2 \times \mathbb{S}^1$, on the other hand, is unstabilized, though it is reducible. It is the unique example of a reducible Heegaard splitting that is not stabilized. You will establish these two facts in the exercises.

Lemma 6.3.4. *An incompressible and boundary incompressible surface in a handlebody is a disk.*

Note that the statement is false if we drop the assumption of boundary incompressibility. (For instance, handlebodies of genus at least one contain incompressible boundary parallel annuli.)

Proof. Let H be a handlebody of genus g. We proceed by induction on g. Note that an incompressible surface in a 3-ball must be a disk. See Example 3.4.5. So suppose $g > 0$. Let S be an incompressible surface in H and let D be a component of a set of meridian disks for H. Then S can be isotoped to be disjoint from D by standard innermost disk and outermost arc arguments. Thus our incompressible and boundary incompressible surface sits in $H \backslash D$, a handlebody of lower genus, where it is still incompressible and boundary incompressible. □

The following theorem is due to Haken. It is one of the fundamental theorems concerning Heegaard splittings. One consequence of this theorem is that a Heegaard splitting of a connected sum of 3-manifolds can be factored into Heegaard splittings of the summands.

Theorem 6.3.5 (Haken)**.** *Suppose M is a reducible 3-manifold and $M = V \cup_S W$ is a Heegaard splitting. Then $M = V \cup_S W$ is reducible.*

Proof. Let $\tilde{S}$ be an essential 2-sphere in M and suppose that $\tilde{S}$ is chosen so that the number of components, $\#|\tilde{S} \cap S|$, of $\tilde{S} \cap S$ is minimal over all essential 2-spheres in M. You will prove in the exercises that handlebodies are irreducible; thus $\#|\tilde{S} \cap S| > 0$. Assuming $\#|\tilde{S} \cap S| > 0$, we reason as follows:

Claim 1. $\tilde{S} \cap V$ *is incompressible in V and $\tilde{S} \cap W$ is incompressible in W.*

Suppose that $\tilde{S} \cap V$, say, is compressible in V. Then there is a simple closed curve $c \in \tilde{S}$ that does not bound a disk in $\tilde{S} \cap V$ but bounds a disk $D \subset V$ that is disjoint from $\tilde{S}$. In particular, $\tilde{S} \backslash c = S_1 \sqcup S_2$. Set $\tilde{S}_i = S_i \cup D \cup c$.

Note that
$$\#|\tilde{S} \cap S| = \#|\tilde{S}_1 \cap S| + \#|\tilde{S}_2 \cap S|.$$
Furthermore, since c does not bound a disk in $\tilde{S} \cap V$,
$$\#|\tilde{S}_1 \cap S| > 0, \qquad \#|\tilde{S}_2 \cap S| > 0.$$

If $\tilde{S}_1$ is inessential (in M), then $\tilde{S}_2$ is isotopic to $\tilde{S}$; see the proof of Lemma 3.2.3. This implies that $\tilde{S}_2$ is essential in M; yet
$$\#|\tilde{S}_2 \cap S| < \#|\tilde{S} \cap S|,$$
so this violates our choice of $\tilde{S}$, chosen so as to minimize $\#|\tilde{S} \cap S|$ over all essential 2-spheres in M. Thus $\tilde{S}_1$ is essential, but
$$\#|\tilde{S}_1 \cap S| < \#|\tilde{S} \cap S|,$$
again violating our choice of $\tilde{S}$. This proves the claim.

By Lemma 6.3.4, $\tilde{S} \cap V$ is either boundary compressible in V or is a disk. Likewise, $\tilde{S} \cap W$ is either boundary compressible in W or is a disk. Suppose that a component, Q, of, say, $\tilde{S} \cap V$ is boundary compressible in V. The boundary compressing disk describes an isotopy of $\tilde{S}$. See Figures 6.12 and 6.13.

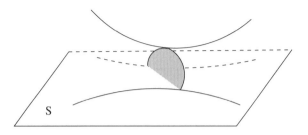

Figure 6.12. *A boundary compressing disk.*

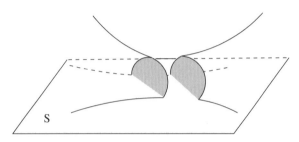

Figure 6.13. *An isotopy through a boundary compressing disk.*

This isotopy has the effect of either producing one component, Q', out of Q, where
$$\chi(Q') = \chi(Q) + 1,$$
or of producing two components, Q_1, Q_2, out of Q, where
$$\chi(Q_1) + \chi(Q_2) = \chi(Q) + 1.$$
See Figures 6.14 and 6.15 to see how the boundary compressing disks meet $\tilde{S} \cap V$. We call the former non-separating and the latter separating.

If we choose our essential 2-sphere in M so that $\chi(\tilde{S} \cap V)$ is maximal, while retaining the minimality of $\#|\tilde{S} \cap S|$ among essential 2-spheres in M, then it follows from the Euler characteristic computation above that $\tilde{S} \cap V$ consists of disks. Under these assumptions, the argument in Claim 1 still shows that components of $\tilde{S} \cap W$ are incompressible. (Indeed, the two spheres created in the proof of Claim 1 would also intersect V in disks.)

Suppose that the number of components/disks in $\tilde{S} \cap V$ is n. Then $\tilde{S} \cap W$ is a planar surface with n boundary components. Lemma 6.3.4 tells us that each component of $\tilde{S} \cap W$ is either boundary compressible or a disk. A non-separating boundary compression corresponds to an arc that has its endpoints on two distinct components of $\partial(\tilde{S} \cap W)$. A separating boundary compression corresponds to an arc that has its endpoints on one component of $\partial(\tilde{S} \cap W)$. Arcs corresponding to successive boundary compressions can be simultaneously embedded in $\tilde{S} \cap W$. Note that there can be at most

6.3. Reducibility and Stabilization

$-\chi(\tilde{S} \cap W)$ non-parallel separating essential arcs in $\tilde{S} \cap W$ corresponding to separating boundary compressions. Thus there can be at most $-\chi(\tilde{S} \cap W) = n - 2$ such arcs.

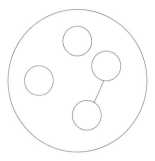

Figure 6.14. *Non-separating.*

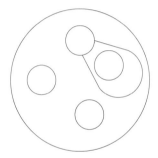

Figure 6.15. *Separating.*

The effect of doing a boundary compression of the non-separating type is to lower the number of components of $\tilde{S} \cap V$ by one and to leave the number of components of $\tilde{S} \cap W$ unchanged. The effect of doing a boundary compression of the separating type is to leave the number of components of $\tilde{S} \cap V$ unchanged and to raise the number of components of $\tilde{S} \cap W$ by one.

We now perform boundary compressions on $\tilde{S} \cap W$ in an effort to produce a 2-sphere $\bar{S}$ isotopic to $\tilde{S}$ such that $\bar{S} \cap V$ is a connected planar surface and $\bar{S} \cap W$ consists of disks. As we do so, we need perform at most $n - 2$ separating boundary compressions. Since $\tilde{S} \cap W$ is connected and since each boundary compression of the separating type increases the number of components of the portion of the 2-sphere lying in W by one, it follows that $\bar{S} \cap W$ has at most $n - 1$ components. In particular,

$$\#|\bar{S} \cap S| = \#|\tilde{S} \cap S| - 1.$$

This contradicts our assumption that $\#|\tilde{S} \cap S|$ is minimal over all essential 2-spheres in M.

It follows that there are no boundary compressions. In particular, $\tilde{S} \cap V$ and $\tilde{S} \cap W$ are disks and $\tilde{S} \cap S$ is an essential simple closed curve. Therefore $M = V \cup_S W$ is reducible. □

Exercises

Exercise 1. Prove that any two Heegaard splittings resulting from elementary stabilizations of a Heegaard splitting $M = V \cup_F W$ are equivalent. (This property is called the *uniqueness of stabilization*.)

Exercise 2. Prove that a stabilized Heegaard splitting of a 3-manifold not equal to $\mathbb{S}^3$ is reducible.

Exercise 3. Prove that a reducible Heegaard splitting of a prime 3-manifold not equal to $\mathbb{S}^2 \times \mathbb{S}^1$ is stabilized.

Exercise 4. Prove that handlebodies are irreducible.

6.4. Waldhausen's Theorem

One of the first theorems proved about Heegaard splittings is due to Waldhausen. See [**155**]. He analyzed the Heegaard splittings of $\mathbb{S}^3$ and found that there is only one Heegaard splitting of $\mathbb{S}^3$ of any given genus. Moreover, the Heegaard splittings of $\mathbb{S}^3$ arise via stabilization of the genus 0 splitting.

Other proofs of this theorem have been given since then; see for instance [**136**]. But Waldhausen's proof remains of interest. The strategy is very natural and may generalize to prove similar results in situations where the techniques of [**136**] fail to apply. This is particularly relevant as recent years have seen a resurgence of interest in issues pertaining to stabilization of Heegaard splittings. Waldhausen's strategy lies in applying the Reidemeister-Singer Theorem to compare a given Heegaard splitting of $\mathbb{S}^3$ to the genus 0 Heegaard splitting. The two collections of stabilizing pairs of disks are compared and played off against each other. Below, we provide a sketch of the argument. We will follow Waldhausen's strategy, but not his terminology.

Definition 6.4.1. Let $M = V \cup_F W$ be a Heegaard splitting. A *good system* of n disks in V is the union of n disjoint disks $v = v_1 \cup \cdots \cup v_n$ in V and n disjoint disks $w = w_1 \cup \cdots \cup w_n$ in W such that:

(1) $\partial v_j \cap \partial w_j$ consists of exactly one point and

(2) $\partial v_i \cap \partial w_j = \emptyset$ when $i > j$.

If, in addition, $\partial v_i \cap \partial w_j = \emptyset$ when $i < j$, then the system of disks is called a *stabilizing* system of disks.

6.4. Waldhausen's Theorem   189

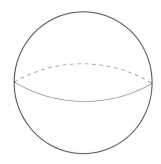

Figure 6.16. *A Heegaard diagram of* $\mathbb{S}^3$.

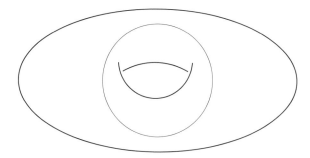

Figure 6.17. *Another Heegaard diagram of* $\mathbb{S}^3$.

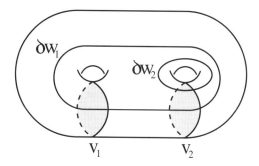

Figure 6.18. *A good system of disks.*

In Figure 6.18, a disk slide of w_1 over w_2 turns a good system of disks into a stabilizing system of disks. This is always possible; that is, for any good system of disks $v \cup w$ there is a sequence of disk slides of components of w over components of w that yields a stabilizing system of disks $v \cup w'$.

Recall that an elementary stabilization of a Heegaard splitting can be expressed via a pairwise connected sum $(M, F) \# (\mathbb{S}^3, T)$, where T is the unknotted torus in $\mathbb{S}^3$. Here T divides $\mathbb{S}^3$ into two solid tori. The meridian of one of these solid tori is a longitude in the other. Since the meridian and longitude of a given solid torus intersect in one point, so do the boundaries of the two meridian disks. This pair of disks persists in the Heegaard splitting obtained by the elementary stabilization.

If a stabilized Heegaard splitting is the result of n elementary stabilizations, then there will be n pairs of disks. Moreover, the pairs will be disjoint from each other, and in each pair, the boundaries of the two disks will intersect exactly once. In other words, the stabilized Heegaard splitting comes with a system of stabilizing disks. Cutting along one disk per pair recreates the Heegaard splitting from which the stabilized Heegaard splitting was obtained.

Remark 6.4.2. In the exercises, you will prove two facts about a good system of disks $v \cup w$ for a Heegaard splitting $M = V \cup_F W$: (1) Cutting V along v yields a handlebody whose complement is also a handlebody. So this yields a Heegaard splitting. Likewise, cutting W along w yields a Heegaard splitting. (2) These two Heegaard splittings are equivalent.

Definition 6.4.3. A Heegaard splitting obtained from the Heegaard splitting $M = V \cup_S W$ by cutting along a component in a good system of disks is called a *destabilization*. If the good system of disks has $2n$ components (n in v and n in w), then cutting along all components in v is called an *n-fold destabilization* (along v).

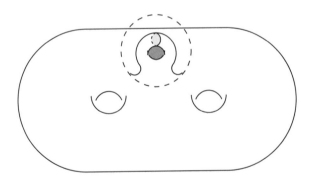

Figure 6.19. *A stabilizing pair of disks.*

Remark 6.4.4. If follows from Remark 6.4.2 and the exercises that if both $v \cup w$ and $v' \cup w$ are good systems of disks, then cutting along v yields the same Heegaard splitting as cutting along v'.

The following lemma sets the stage for the proof of Waldhausen's Theorem. Among other things, it establishes that, given two good systems of disks, a sequence of slides of components of one over components of the other guarantees a certain disjointness.

6.4. Waldhausen's Theorem

Lemma 6.4.5. *If $M = V \cup_F W$ and $M = V' \cup_{F'} W'$ are Heegaard splittings, then there is a Heegaard splitting $M = X \cup_S Y$ and good systems $x \cup y$ and $x' \cup y'$ of disks such that the following hold:*

(1) $M = V \cup_F W$ *is an n-fold destabilization of* $M = X \cup_S Y$ *along x;*

(2) $M = V' \cup_{F'} W'$ *is an m-fold destabilization of* $M = X \cup_S Y$ *along x';*

(3) $x \cap x' = \emptyset$;

(4) $y \cap y' = \emptyset$.

Proof. By the Reidemeister-Singer Theorem, $M = V \cup_F W$ and $M = V' \cup_{F'} W'$ have a common stabilization. We denote this common stabilization by $M = X \cup_S Y$. It remains to show that $M = X \cup_S Y$ satisfies the required properties. Properties (1) and (2) follow immediately from the definition of stabilization.

To show that properties (3) and (4) are satisfied, it suffices to show that we may alter x, y, x', y' so that $x \cap x' = \emptyset$ and $y \cap y' = \emptyset$. We first consider closed components of intersection and note that these may be removed via an innermost disk argument. Next we consider an arc of intersection, say between $x_i \subset x$ and $x'_j \subset x'$. Furthermore, we assume that this arc is outermost in x'_j. The outermost arc cuts off a disk D from x'_j. Cutting x_i along $\partial D \cap x_i$ and attaching copies of D to the resulting boundary arcs yields two disks. One of these contains the point of intersection between x_i and y_i. Denote this latter disk by $\tilde{x}_i$. We replace x_i with $\tilde{x}_i$. By Remark 6.4.4, the veracity of properties (1) and (2) is unaffected, but the number of arcs of intersection between $x \cap x'$ has been reduced. □

Below you will find a sketch of Waldhausen's argument. Many (re)proofs of Waldhausen's Theorem have been given. The proof below follows Waldhausen's original proof closely. See [**139**] for background information. A proof that deserves special mention is that given by Scharlemann and Thompson in [**136**].

Theorem 6.4.6 (Waldhausen). *The 3-sphere has a unique Heegaard splitting of any given genus.*

Proof. It follows from the Schönflies Theorem that $\mathbb{S}^3$ has a unique Heegaard splitting of genus 0. Let $\mathbb{S}^3 = V \cup_F W$ be a Heegaard splitting of genus greater than 0. We wish to show that $\mathbb{S}^3 = V \cup_F W$ is stabilized. The theorem will then follow from the uniqueness of stabilization.

Choose a Heegaard splitting $\mathbb{S}^3 = X \cup_S Y$ as in Lemma 6.4.5 such that both $\mathbb{S}^3 = V \cup_F W$ and the genus 0 Heegaard splitting $\mathbb{S}^3 = V' \cup_{F'} W'$ are destabilizations of $\mathbb{S}^3 = X \cup_S Y$. As above, we denote the corresponding systems of disks by $x \cup y$ and $x' \cup y'$. We will also assume that $\mathbb{S}^3 = X \cup_S Y$ has been chosen to be minimal genus subject to these conditions.

Case 1. $x_n \cap y' = \emptyset$.

Then ∂x_n survives in the boundary of the result of cutting Y along y'. But the result of cutting Y along y' is a 3-ball B. Thus $\partial x_n \subset \partial B$ bounds a disk in this 3-ball. Since this 3-ball is entirely contained in Y, it follows that ∂x_n bounds a disk D_y in Y. Thus $x_n \cup D_y$ forms a 2-sphere in $\mathbb{S}^3$. By Alexander's Theorem, this 2-sphere is separating. It follows that ∂x_n is separating on S. This is a contradiction since ∂x_n intersects ∂y_n exactly once.

Case 2. $x_n \cap y' \neq \emptyset$ and consists of exactly one point.

We may assume that the point of intersection lies in, say, y'_j. Then cutting Y along y'_j yields a Heegaard splitting isotopic to the Heegaard splittings obtained by cutting X along x_n or x'_j, respectively (see Remark 6.4.4). It follows that this resulting Heegaard splitting is a common stabilization of both $\mathbb{S}^3 = V \cup_F W$ and the genus 0 Heegaard splitting of $\mathbb{S}^3$. This contradicts the minimality assumption on the genus of $\mathbb{S}^3 = X \cup_S Y$.

Case 3. $x_n \cap y' \neq \emptyset$ and consists of more than one point but no more than one point in any one component of y.

In this case we may alter y' by disk slides to reduce the number of points in $x_n \cap y' \neq \emptyset$.

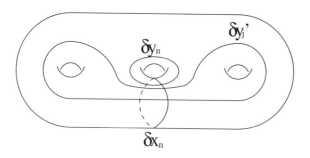

Figure 6.20. Setting for a disk slide of y'_j over y_n.

Case 4. x_n meets a component of y' in more than one point.

We may assume that x_n meets, say, y'_k, at least two times. This allows us to alter y' by a disk slide over y_n to reduce the number of points of intersection between x_n and y'. See Figure 6.20. □

Exercises

Exercise 1. Show that given a good system of disks $v \cup w$, there is a stabilizing system of disks $v \cup w'$, where w' is obtained from w by disk slides.

Exercise 2. Suppose $v \cup w$ is a good system of disks for the Heegaard splitting $M = V \cup_S W$. Show that cutting V along v yields the same Heegaard splitting as cutting along w.

6.5. Structural Theorems

Waldhausen's Theorem concerning Heegaard splittings of $\mathbb{S}^3$ makes a statement about the structure of all Heegaard splittings of $\mathbb{S}^3$. It is the first of a sequence of theorems describing the structure of Heegaard splittings of 3-manifolds and classes of 3-manifolds. F. Bonahon and J.-P. Otal proved the following:

Theorem 6.5.1 (Bonahon-Otal). *All Heegaard splittings of a lens space $M \neq \mathbb{S}^3$ are stabilizations of a unique genus 1 Heegaard splitting of M.*

For a proof of Theorem 6.5.1, see [14].

Definition 6.5.2. The *genus* of a 3-manifold M is the smallest possible genus of a Heegaard splitting for M.

Recall that lens spaces are obtained by identifying the boundaries of two solid tori. This is a description of lens spaces in terms of Heegaard splittings. This description makes it clear that the genus of a lens space is 1. (It can't be 0 since you showed above that only the 3-sphere has genus 0.) The theorem of Bonahon and Otal tells us that each lens space has a unique Heegaard splitting of genus 1 and every other of its Heegaard splittings is a stabilization of this Heegaard splitting of genus 1.

Theorem 6.5.3 (Boileau-Otal). *All Heegaard splittings of the 3-torus are stabilizations of the genus 3 Heegaard splitting described above.*

For a proof of Theorem 6.5.3, see [13]. The theorem of Boileau and Otal explicitly describes the structure of Heegaard splittings of the 3-torus. The construction can be generalized to manifolds homeomorphic to (closed orientable surface) $\times \mathbb{S}^1$.

Definition 6.5.4. Let Q be a closed orientable surface. Let p be a point in Q and let $a_1, \ldots, a_{2g}$ be a collection of arcs based at p that cut Q into a disk. Let t be a point in $\mathbb{S}^1$. Denote a closed regular neighborhood of $a_1 \times t \cup \cdots \cup a_{2g} \times t \cup p \times \mathbb{S}^1$ by V. Denote the closure of the complement of

V by W. Set $S = \partial V$ and $M = Q \times \mathbb{S}^1$. You will show in the exercises that $M = V \cup_S W$ is a Heegaard splitting. This Heegaard splitting is called the *standard Heegaard splitting* of M.

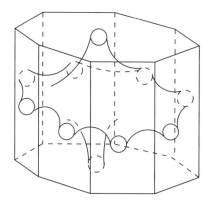

Figure 6.21. *The standard Heegaard splitting of $Q \times \mathbb{S}^1$.*

Theorem 6.5.5 (Schultens). *All Heegaard splittings of (closed orientable surface) $\times \mathbb{S}^1$ are stabilizations of the standard Heegaard splitting.*

For a proof of Theorem 6.5.5, see [**143**]. The standard Heegaard splitting of (closed orientable surface) $\times \mathbb{S}^1$ is described in terms of a *spine* for V, i.e., a graph whose regular neighborhood is V. This is a useful strategy in many settings. Consider the following example:

Definition 6.5.6. Let M be a prism manifold, i.e., a Seifert fibered space with base orbifold the 2-sphere with three exceptional points a, b, c. Denote the base orbifold of M by O. Let γ be an arc in M whose projection is a simple arc connecting the exceptional points a and b. Denote γ together with the exceptional fibers that project to a and b by Γ. See Figure 6.22. Denote a closed regular neighborhood of Γ by V, the closure of the complement of V by W and the surface $V \cap W$ by S. You will show in the exercises that $M = V \cup_S W$ is a Heegaard splitting for the prism manifold M.

Any Heegaard splitting of a prism manifold of this form is called a *vertical* Heegaard splitting.

The notion of a vertical Heegaard splitting can be generalized to more complicated Seifert fibered spaces. This can then be seen as the canonical construction of Heegaard splittings for Seifert fibered spaces. There is a less canonical construction that yields Heegaard splittings for some, but not all, Seifert fibered spaces. We will not describe it here. Suffice it to say that this type of Heegaard splitting is called a *horizontal* Heegaard splitting. It

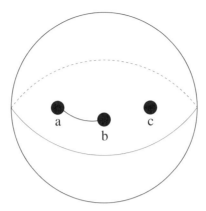

Figure 6.22. *A vertical Heegaard splitting of a prism manifold.*

was proved by Y. Moriah and the author that under certain orientability assumptions, all Heegaard splittings of Seifert fibered spaces are horizontal or vertical. See [**106**].

Another structural theorem for Heegaard splittings for a class of manifolds is that of D. Cooper and M. Scharlemann for solvmanifolds; see [**29**]. A solvmanifold is a 3-manifold that is the mapping torus of an automorphism ϕ : torus $\to$ torus such that the absolute value of the trace of (the matrix corresponding to) ϕ is strictly greater than 2.

A solvmanifold possesses a standard Heegaard splitting that is constructed analogously to the standard Heegaard splitting of the product manifolds above. It is of genus 3. Typically, this is the unique unstabilized Heegaard splitting of this manifold. However, in those cases where (the matrix corresponding to) ϕ is conjugate to a matrix of the form

$$\begin{pmatrix} \pm m & -1 \\ 1 & 0 \end{pmatrix}, \quad m \geq 3,$$

the standard Heegaard splitting of genus 3 is stabilized. In this case there is a genus 2 Heegaard splitting that is the unique unstabilized Heegaard splitting of the solvmanifold.

Exercises

Exercise 1. Show that the splitting described in Definition 6.5.4 is indeed a Heegaard splitting.

Exercise 2. The standard Heegaard splitting of (closed orientable surface) $\times \, \mathbb{S}^1$ was described in terms of a spine for V. Describe a spine for W.

Exercise 3. Show that the splitting described in Definition 6.5.6 is indeed a Heegaard splitting.

Exercise 4. The standard Heegaard splitting of a prism manifold was described in terms of a spine for V. Describe a spine for W.

6.6. The Rubinstein-Scharlemann Graphic

In the 1990s Rubinstein and Scharlemann decided to investigate the relative positioning of two Heegaard splittings of the same manifold with respect to each other. To do so, they reinterpreted and reformulated the work of Cerf. The specific tool they used has become known as the Rubinstein-Scharlemann graphic. See [**130**], [**129**], and [**131**].

Most important here is the fact that the handlebody V is a regular neighborhood of the spine Γ. This means that $V \backslash \Gamma$ is homeomorphic to $\partial V \times [0,1)$. Consequently, if $M = V \cup_S W$ is a Heegaard splitting with Γ_v a spine of V and Γ_w a spine of W, then $M \backslash (\Gamma_v \cup \Gamma_w)$ is homeomorphic to $S \times (-1,1)$. Furthermore, the image of $S \times \{t\}$ under this homeomorphism is isotopic to S, for all $t \in (-1,1)$, and hence defines the same Heegaard splitting as S. In this section we denote the interval $[-1,1]$ by I.

Definition 6.6.1. Let $M = V \cup_S W$, Γ_v, and Γ_w be as above. A continuous map $h : S \times (I, \partial I) \to (M, \Gamma_v \cup \Gamma_w)$ that is a homeomorphism on $S \times (-1,1)$ and maps $S \times \partial\{-1\}$ to Γ_v and $S \times \partial\{1\}$ to Γ_w is called a *sweepout* of M. We denote the image of (S, t) be S_t.

Suppose that $M = X \cup_Q Y$ is also a Heegaard splitting. Then for Γ_x a spine of X and Γ_y a spine of Y there is also a sweepout $g : Q \times (I, \partial I) \to (M, \Gamma_x \cup \Gamma_y)$. Rubinstein and Scharlemann were interested in analyzing the intersections of Q_r and S_t.

Suppose that the two sweepouts are transverse. We consider $(r,t) \in I \times I$. If (r,t) is generic, then $Q_r \pitchfork S_t$. As it turns out (for details see [**125**]), there are four types of points in the interior of the square $I \times I = \{(r,t) : 0 \le r \le 1, 0 \le t \le 1\}$:

(1) those where Q_r and S_t meet transversely;

(2) those where Q_r and S_t meet transversely except at a single non-degenerate tangent point (i.e., a point of tangency modeled on the point of tangency of $z = x^2 \pm y^2$ and $z = 0$);

(3) those where Q_r and S_t meet transversely except at two non-degenerate tangent points;

(4) those where Q_r and S_t meet transversely except at a single degenerate critical point modeled on $Q_r = \{(x,y,z)|z = 0\}$ and $S_t = \{(x,y,z)|z = x^2 + y^3\}$.

The last two types of points are isolated; the third type is called *crossing vertices*, the fourth *birth-death vertices*. The points of the second type occur in codimension 1 strata that we call *edges*. The points of the first type make up the rest of the interior of the square and thus constitute connected open sets that we call *regions*. The edges and vertices make up a graph that is called the *Rubinstein-Scharlemann graphic* and is denoted by Γ. The Rubinstein-Scharlemann graphic naturally extends to the closed square, where we continue to call it the Rubinstein-Scharlemann graphic.

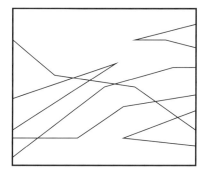

Figure 6.23. *The Rubinstein-Scharlemann graphic.*

Consider a point $(r,t) \in (I \times I) \setminus \Gamma$ and denote the region it lies in by R. The corresponding surfaces Q_r and S_t intersect transversely. We are interested in the curves of intersection that are essential in Q_r. If there is such a curve that bounds a disk in X (or Y, respectively), then we label the region R with an X (or Y, respectively). We are also interested in the curves of intersection that are essential in S_t. If there is such a curve that bounds a disk in V (or W, respectively), then we label the region R with a V (or W, respectively).

If we imagine a path within a region, then each point on this path corresponds to a pair of surfaces that intersect transversely. Moreover, since all these pairs of surfaces intersect transversely, their curves of intersection are remaining constant up to isotopy. In particular, the labeling is well-defined on the region!

Our labeling has many consequences. The lemma below follows immediately from these definitions.

Lemma 6.6.2. *If any region is labeled both X and Y, then $M = X \cup_Q Y$ is weakly reducible. Analogously, if any region is labeled both V and W, then $M = V \cup_S W$ is weakly reducible.*

Part of the strength of this labeling scheme is that we can say something about the labeling of regions that are adjacent along an edge. To this end, we imagine a short path in $I \times I$ that connects two such regions. We are especially interested in the point at which the path crosses the edge. See Figures 6.24 and 6.25.

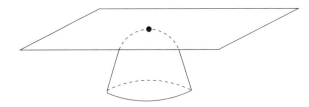

Figure 6.24. *Surfaces corresponding to an edge in the Rubinstein-Scharlemann graphic.*

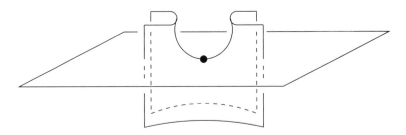

Figure 6.25. *Surfaces corresponding to an edge in the Rubinstein-Scharlemann graphic.*

If the path crosses the edge in a point corresponding to surfaces meeting as in Figure 6.24, then only inessential curves of intersection are created or destroyed. From the point of view of our labeling, this is of no interest at all. On the other hand, if the path crosses the edge in a point corresponding to surfaces meeting as in Figure 6.25, then essential curves of intersection may be created or destroyed. See Figure 6.26.

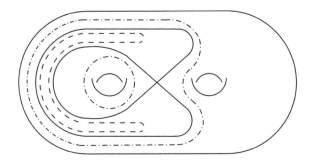

Figure 6.26. *Possible effect of crossing and edge.*

6.6. The Rubinstein-Scharlemann Graphic

Lemma 6.6.3. *Suppose that regions R_1 and R_2 are adjacent along an edge. Suppose further that R_1 is labeled X and R_2 is labeled Y. Then $M = X \cup_Q Y$ is weakly reducible.*

Proof. If R_1 is also labeled Y or if R_2 is also labeled X, then the result follows from Lemma 6.6.2. So we will assume that R_1 is not labeled Y and R_2 is not labeled X.

Consider a short path in $I \times I$ that begins in R_1 and ends in R_2 and intersects the edge between these two regions in a single point. The point of intersection (r,t) corresponds to a pair of surfaces Q_r, S_t such that $Q_r \cap S_t$ contains a figure 8 (cf. Figure 6.25). If we think of Q_r as the splitting surface Q and let t vary (exhibiting the horizontal surfaces S_t stacked neatly on top of each other), then to one side (say in $Q \cap S_{t-\epsilon}$) of this figure 8 there is an essential curve that bounds a disk in X. To the other side (in $Q \cap S_{t+\epsilon}$) of this figure 8 there is an essential curve that bounds a disk in Y. Thus $M = X \cup_Q Y$ is weakly reducible. $\square$

More technical arguments can be employed to say more about the labeling of the regions. The following results deserve particular mention:

Lemma 6.6.4. *If all four labelings X, Y, V, W appear in the quadrants of a crossing vertex, then either two opposite quadrants are unlabeled or one of $M = X \cup_Q Y$ or $M = V \cup_S W$ is weakly reducible or $M = \mathbb{S}^3$.*

Lemma 6.6.5. *There is an unlabeled region.*

Proposition 6.6.6. *For one of the pairs of labels X, Y or V, W there is a (generic) path that traverses only unlabeled regions and begins at an edge labeled X (or V, respectively) and ends at an edge labeled Y (or W, respectively).*

One of the most impressive results obtained by Rubinstein and Scharlemann using this general setup is the following:

Theorem 6.6.7. *Suppose $M = X \cup_Q Y$ and $M = V \cup_S W$ are strongly irreducible Heegaard splittings of genus q and s, respectively. Suppose also that $q \leq s$. Then there is a genus $8q + 5s - 11$ Heegaard splitting of M that is a stabilization of both $M = X \cup_Q Y$ and $M = V \cup_S W$.*

Using the additional machinery of generalized Heegaard splittings, described in subsequent sections, Rubinstein and Scharlemann obtained a quadratic bound on the number of stabilizations required in a generalized version of Theorem 6.6.7 to the case where the Heegaard splittings are not strongly irreducible.

Exercises

Exercise 1. Prove Lemma 6.6.5.

Exercise 2. Let M be a prism manifold, i.e., a Seifert fibered space that is fibered over the sphere and has three exceptional fibers. Construct a spine for a handlebody V by connecting the three exceptional fibers together by arcs that project to embedded arcs. The closure, W, of the complement of this handlebody is also a handlebody. Show that the Heegaard splitting $M = V \cup_S W$ thus defined is stabilized.

Exercise 3. Prism manifolds have up to three distinct vertical Heegaard splittings. Show that these become isotopic after one stabilization. (Hint: See the exercise above.)

6.7. Weak Reducibility and Incompressible Surfaces

In this section we prove a theorem of Casson and Gordon; see [**24**]. The theorem establishes a connection between a Heegaard splitting being weakly reducible and the existence of an incompressible surface. This theorem proved to be seminal. Eventually, it engendered the concept of a thin manifold decomposition of a 3-manifold pioneered by Scharlemann and Thompson. This concept in turn gave rise to the notion of a generalized strongly irreducible Heegaard splitting, a structure possessed by every compact 3-manifold.

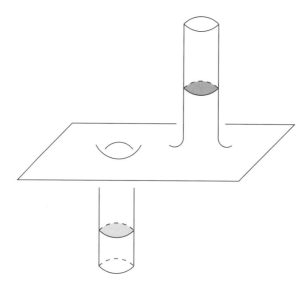

Figure 6.27. *Local picture of a portion of a weakly reducible Heegaard splitting.*

6.7. Weak Reducibility and Incompressible Surfaces

Theorem 6.7.1. *Suppose M is a closed orientable 3-manifold and $M = V \cup_S W$ is a weakly reducible Heegaard splitting. Then either M contains an incompressible surface or $M = V \cup_S W$ is reducible.*

Proof. Let $\mathcal{D}$ be a non-empty disjoint union of non-parallel essential disks in V and let $\mathcal{E}$ be a non-empty disjoint union of non-parallel essential disks in W such that $\partial \mathcal{D} \cap \partial \mathcal{E} = \emptyset$. Consider $S \backslash (\mathcal{D} \cup E)$. For each component of this surface, take the union with appropriate components of $\mathcal{D} \cup E$ to obtain a closed surface S^*. After a small isotopy, S^* is embedded in M. (For every disk D in $\mathcal{D} \cup E$, there will be two remnants of D in S^*.)

Since $M = V \cup_S W$ is weakly reducible, we can choose $\mathcal{D}$ and $\mathcal{E}$ to be non-empty. More importantly, we will assume, in what follows, that $\mathcal{D} \cup E$ is chosen so that $\chi(S^*)$ is maximal.

Case 1. A component, Q, of S^* has positive genus.

In this case it follows from our maximality assumption that Q is incompressible.

Case 2. All components of S^* are 2-spheres.

Let $\mathcal{V}$ be the components of S^* that meet the interior of V and let $\mathcal{W}$ be the components of S^* that meet the interior of W.

Claim 1. $\mathcal{V} \cap \mathcal{W} \neq \emptyset$.

If $\mathcal{V} \cap \mathcal{W} = \emptyset$, then reversing the cut and paste operations performed above would connect components in $\mathcal{V}$ with components in $\mathcal{V}$ and components in $\mathcal{W}$ with components in $\mathcal{W}$ and this would result in at least two components. Since S is connected, this is impossible.

Let $\tilde{S}$ be a component of $\mathcal{V} \cap \mathcal{W}$. Then $\tilde{S}$ lies mostly in S. Furthermore, $\tilde{S} \cap V$ and $\tilde{S} \cap W$ are non-empty and consist of disks. Let c be a simple closed curve in $\tilde{S}$ that separates the components of $\tilde{S} \cap V$ from the components of $\tilde{S} \cap W$. Note that c is also a simple closed curve in S.

Claim 2. c is an essential curve in S.

There are essential curves to either side of c; thus c can't bound a disk in S.

Now the disk in $\tilde{S} \backslash c$ that meets V can be isotoped slightly to one side of S to lie entirely in V and the other can be similarly isotoped to lie entirely in W. This shows that $M = V \cup_S W$ is reducible. $\square$

In the case in which $M = V \cup_S W$ is reducible, two things can happen: (1) The 2-sphere constructed can be essential; (2) the 2-sphere constructed can be inessential. In the first case, M contains an incompressible surface, namely the 2-sphere. In the second case, the 2-sphere splits off an $\mathbb{S}^3$ summand with a Heegaard splitting of positive genus. It then follows from the theorem of Waldhausen discussed in Section 6.4 that $M = V \cup_S W$ is in fact stabilized.

So far, we have discussed Heegaard splittings in the context of closed 3-manifolds. The notion can be generalized in more than one way. The most common such generalization involves the notion defined below.

Definition 6.7.2. A *compression body* is a compact 3-manifold W that can be obtained from a closed surface Q and 0-handles by attaching 1-handles that don't meet $Q \times \{0\} \subset Q \times [0, 1]$.

We denote $Q \times \{0\}$ by $\partial_- W$ and $\partial W \backslash \partial_- W$ by $\partial_+ W$.

Note that handlebodies form a subset of compression bodies.

Definition 6.7.3. Let M be a compact 3-manifold. A *Heegaard splitting* of M is a decomposition $M = V \cup_S W$, where V, W are compression bodies and $S = \partial_+ V = \partial_+ W$.

The terminology introduced for Heegaard splittings of closed 3-manifolds (equivalence, genus, etc.) carries over to this more general setting.

Exercises

Exercise 1. Consider the standard genus 3 Heegaard splitting of the 3-torus. Show that it is weakly reducible by exhibiting a pair of disks that satisfy the definition. Then use these disks as in Theorem 6.7.1 to produce an incompressible surface.

Exercise 2. Generalize Theorem 6.7.1 to manifolds with boundary.

6.8. Generalized Heegaard Splittings

Section 6.6 provided a taste of the utility of strongly irreducible Heegaard splittings. Another crucial feature of strongly irreducible Heegaard splittings revolves around the fact that their splitting surfaces behave like incompressible surfaces in many ways. Recall, for instance, that two incompressible surfaces can be isotoped to intersect only in curves that are essential in both surfaces. Analogously:

Lemma 6.8.1. *Let M be an irreducible 3-manifold. Let F be an incompressible surface (possibly with boundary) and let $M = V \cup_S W$ be a strongly irreducible Heegaard splitting. Then S can be isotoped so that all components of $F \cap S$ are essential in both F and S.*

6.8. Generalized Heegaard Splittings

Proof. Let Γ_v be a spine of V and let Γ_w be a spine of W. Let h be a sweepout of $M = V \cup_S W$. For t close to -1, $S_t \cap F$ consists of small simple closed curves bounding regular neighborhoods of the points $\Gamma_v \cap F$, that is, disks in V. Note that these curves of intersection are essential in S_t. Likewise, for t close to 1, $S_t \cap F$ consists of small simple closed curves bounding regular neighborhoods of the points $\Gamma_w \cap F$, that is, disks in W. Note that these curves of intersection are essential in S_t.

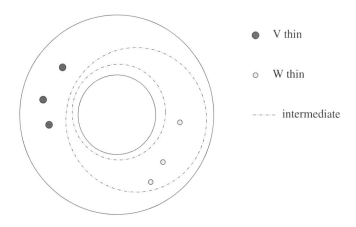

Figure 6.28. $S_t \cap F$, for three different values of t.

The subset of $[-1, 1]$ consisting of t such that $S_t \cap F$ contains simple closed curves that are essential in S_t and bound disks in $V_t \cap F$ is closed, as is the subset of $[-1, 1]$ consisting of t such that $S_t \cap F$ contains simple closed curves that are essential in S_t and bound disks in $W_t \cap F$. Thus these subsets either overlap or are disjoint. Since $M = V \cup_S W$ is strongly irreducible, they can't overlap. Thus they are disjoint. Therefore there exist t such that $S_t \cap F$ contains no simple closed curves that are essential in S_t and bound disks in F. (See Figure 6.28.) Simple closed curves in $S_t \cap F$ that are inessential in S_t can be removed via Lemma 3.2.3. □

Not every 3-manifold possesses a strongly irreducible Heegaard splitting. Consider, for instance, the 3-torus. By Theorem 6.5.3 of Boileau and Otal, it has a unique unstabilized Heegaard splitting. We have seen that this genus 3 Heegaard splitting is weakly reducible. See Figure 6.29.

Heegaard splittings correspond to handle decompositions of a 3-manifold. But not every handle decomposition of a 3-manifold corresponds to a Heegaard splitting. This is because in a handle decomposition, 1-handles need not be attached prior to 2-handles.

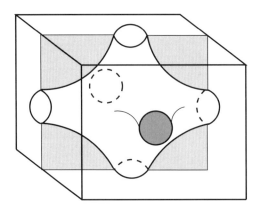

Figure 6.29. *The standard Heegaard splitting of the 3-torus is weakly reducible.*

One motivation for requiring the 1-handles to be attached prior to 2-handles in a Heegaard splitting is to attain a certain symmetry: For a handle decomposition of a closed 3-manifold, a surface, the Heegaard surface, separates the 1-handles from the 2-handles. Since 1-handles are dual to 2-handles, the submanifolds on either side of the surface are handlebodies of the same genus and hence homeomorphic.

Breaking symmetry, we drop the requirement of attaching all 1-handles prior to all 2-handles. Instead of having a single surface that captures all relevant information concerning our splitting, we now have several surfaces deserving attention. More specifically, we attach some number of 1-handles, followed by some number of 2-handles, then more 1-handles, then more 2-handles, and so on. We are now interested in the surface bounding the result of attaching the first collection of 1-handles, the surface bounding the result of attaching the first collection of 2-handles, that bounding the result of attaching the second collection of 1-handles, and so on. See Figure 6.30. Note that successive surfaces of the type described cobound compression bodies. This point of view lies at the heart of Scharlemann and Thompson's notion of thin position for 3-manifolds; see [**137**].

Definition 6.8.2. A *generalized Heegaard splitting* is a decomposition

$$M = (V_1 \cup_{S_1} W_1) \cup_{F_1} (V_2 \cup_{S_2} W_2) \cup_{F_2} \cdots \cup_{F_{n-1}} (V_n \cup_{S_n} W_n),$$

where each V_i and each W_i is a compression body,

$$S_i = \partial_+ V_i = \partial_+ W_i,$$

and

$$F_i = \partial_- W_i = \partial_- V_{i+1}.$$

Two generalized Heegaard splittings

$$M = (V_1 \cup_{S_1} W_1) \cup_{F_1} (V_2 \cup_{S_2} W_2) \cup_{F_2} \cdots \cup_{F_{n-1}} (V_n \cup_{S_n} W_n)$$

6.8. Generalized Heegaard Splittings

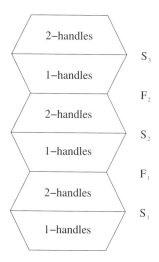

Figure 6.30. *Schematic for a generalized Heegaard splitting.*

and
$$M = (V'_1 \cup_{S'_1} W'_1) \cup_{F'_1} (V'_2 \cup_{S'_2} W'_2) \cup_{F'_2} \cdots \cup_{F'_{n-1}} (V'_n \cup_{S'_n} W'_n)$$
are considered *equivalent* if the collections of surfaces
$$S_1 \cup F_1 \cup S_2 \cup F_2 \cdots \cup F_{n-1} \cup S_n$$
and
$$S'_1 \cup F'_1 \cup S'_2 \cup F'_2 \cdots \cup F'_{n-1} \cup S'_n$$
are isotopic.

The schematic diagram in Figure 6.30 can be misleading in terms of connectedness of the surfaces S_i and F_i. We do not, in fact, require these surfaces to be connected. However, it is common to assume that each V_i and each W_i has only one "active component", i.e., only one component that is not a trivial compression body.

The diagram in Figure 6.31 provides a schematic of a generalized Heegaard splitting that indicates the number of components of each S_i and F_i.

Concerning this topic, a series of lecture notes by T. Saito, M. Scharlemann, and the author is in progress.

Remark 6.8.3. A Heegaard splitting is also a generalized Heegaard splitting, so every 3-manifold possesses a generalized Heegaard splitting.

Definition 6.8.4. A generalized Heegaard splitting
$$M = (V_1 \cup_{S_1} W_1) \cup_{F_1} (V_2 \cup_{S_2} W_2) \cup_{F_2} \cdots \cup_{F_{n-1}} (V_n \cup_{S_n} W_n)$$
is *strongly irreducible* if every F_i is incompressible in M and every $V_i \cup_{S_i} W_i$ is a strongly irreducible Heegaard splitting of $M_i = V_i \cup W_i$.

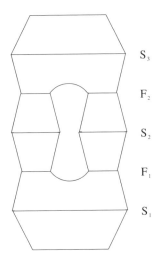

Figure 6.31. *Schematic for a generalized Heegaard splitting.*

Theorem 6.8.5. *Every 3-manifold possesses a strongly irreducible generalized Heegaard splitting.*

Sketch of proof. Let M be a 3-manifold. By the theorem of Moise, M possesses a Heegaard splitting $M = V \cup_S W$. If this Heegaard splitting is strongly irreducible, then there is nothing to prove. If it is weakly irreducible, then there are essential disks $D_v \subset V$ and $D_w \subset W$ that are disjoint. Now D_v is the cocore of a 1-handle and D_w is the core of a 2-handle. It is a deep fact, one we will not address here, that the disjointness of D_v and D_w implies that the 2-handle corresponding to D_w can be "lowered" so as to be attached before the 1-handle corresponding to D_v.

Assume, then, that our handle decomposition is constructed by attaching 1-handles only until one of the 2-handles can be attached; continuing by attaching as many 2-handles as possible; attaching more 1-handles but only until the next 2-handle can be attached; and so forth. You will show in the exercises that $V_i \cup_{S_i} W_i$ must be strongly irreducible. (This completes our sketch of the proof of Theorem 6.8.5.) □

Definition 6.8.6. A rearrangement of the order of attachment of the 1-handles and 2-handles in a Heegaard splitting that results in a strongly irreducible generalized Heegaard splitting is called an *untelescoping*.

Example 6.8.7. Consider the Heegaard splitting of the 3-torus described in Section 6.1. It has genus 3. It is obtained by attaching three 1-handles and then three 2-handles. The order of attachment can be rearranged: We first attach two 1-handles. This gives us V_1. See Figure 6.32.

6.8. Generalized Heegaard Splittings

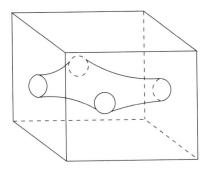

Figure 6.32. V_1.

We then attach one 2-handle. This gives us $V_1 \cup_{S_1} W_1$. The bounding surface of $V_1 \cup_{S_1} W_1$, F_1, consists of two parallel tori. See Figure 6.33.

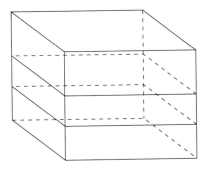

Figure 6.33. F_1.

We have filled up half of the 3-torus. To fill up the remainder we proceed along dual lines: The third 1-handle is attached (a spine of V_2 is depicted in Figure 6.34) and then the remaining two 2-handles.

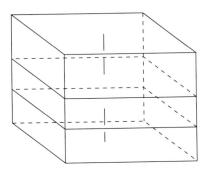

Figure 6.34. A spine of V_2.

Exercises

Exercise 1. Let V be a compression body. Prove that $\partial_- V$ is incompressible in V.

Exercise 2. Describe (in vague terms, invoking the deep fact mentioned in the sketch of a proof of Theorem 6.8.5) why the assumption about the order of attachment of 1-handles and 2-handles in the sketch of a proof of Theorem 6.8.5 forces the Heegaard splittings $V_i \cup_{S_i} W_i$ to be strongly irreducible. See [136].

6.9. An Application

One application of the notion of a generalized Heegaard splitting relates to the behavior of the tunnel number of a knot under the operation of connected sum.

Definition 6.9.1. A *tunnel system* for a knot K in $\mathbb{S}^3$ is a collection of simple arcs $\alpha_1 \cup \cdots \cup \alpha_n$ in $C(K) = \mathbb{S}^3 \backslash \eta(K)$ such that $C(K) \backslash \eta(\alpha_1 \cup \cdots \cup \alpha_n)$ is a handlebody. The *tunnel number* of K, $t(K)$, is the minimum number of components required for a tunnel system.

The tunnel number of a knot is closely related to the Heegaard genus of the knot complement: A tunnel system defines a Heegaard splitting $C(K) = V \cup_S W$ by setting $W = C(K) \backslash \eta(\alpha_1 \cup \cdots \cup \alpha_n)$ and setting V to be the closure of $\eta(\partial C(K)) \cup \eta(\alpha_1 \cup \cdots \cup \alpha_n)$. The genus of this Heegaard splitting is $t(K) + 1$. Conversely, given a Heegaard splitting, we obtain a tunnel system by appropriately manipulating a spine of the compression body containing $\partial C(K)$. It then follows that the genus of $C(K)$ is exactly $t(K) + 1$.

As it turns out, the tunnel number of a knot behaves rather erratically under the operation of connected sum of knots. Morimoto, Sakuma, and Yokota exhibited examples of two knots with tunnel number 1 whose sum has tunnel number 3; see [108]. Morimoto exhibited examples of knots having tunnel numbers 1 and 2, respectively, whose sum has tunnel number 2; see [107]. Kobayashi extended these examples to show that the difference $t(K_1 \# K_2) - t(K_1) - t(K_2)$ can be arbitrarily large; see [81]. The following theorem, due to Scharlemann and the author (see [135]), thus came as a surprise:

Theorem 6.9.2.
$$t(K_1 \# K_2) \geq \frac{2}{5}(t(K_1) + t(K_2)).$$

We will not discuss the proof here, as it is rather technical. But we will discuss the key ingredients of the proof of a related theorem, also by

6.9. An Application

Scharlemann and the author (see [**134**]):

Theorem 6.9.3.
$$t(K_1 \# \cdots \# K_n) \geq n.$$

We observed above that for a knot K to have tunnel number at least n, $C(K)$ must have Heegaard genus at least $n+1$. We prove a specialized form of Theorem 6.9.3:

Proposition 6.9.4. *Let $K = K_1 \# \cdots \# K_n$. (I.e., K is a connected sum of n non-trivial knots.) Suppose that $C(K) = V \cup_S W$ is a strongly irreducible Heegaard splitting of the complement of K. Then the genus of $C(K) = V \cup_S W$ is at least $n+1$.*

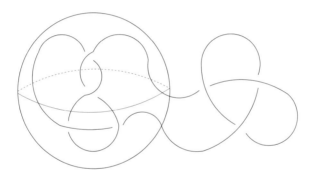

Figure 6.35. *A decomposing sphere.*

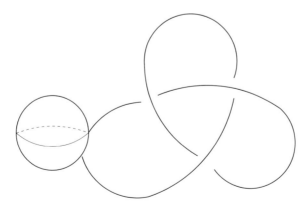

Figure 6.36. *Visualizing the knot complement.*

Proof. First a general fact about a connected sum of n knots: The complement of such a knot will contain a collection of $n-1$ *decomposing spheres*, $\mathcal{S} = S_1 \sqcup \cdots \sqcup S_{n-1}$, characterized by the fact that

$$C(K) \backslash \mathcal{S}$$

consists of punctured copies of $C(K_1) \sqcup \cdots \sqcup C(K_n)$. Each of the spheres intersects K in two points and $C(K)$ in an annulus called a *decomposing annulus*. The decomposing annuli cut $C(K)$ into $\bigcup_{i=1}^{n} C(K_i)$. See Figures 6.35 and 6.36.

Because $C(K) = V \cup_S W$ is strongly irreducible, S can be isotoped so that it intersects each decomposing annulus in curves that are essential in both S and in the decomposing annulus. Let A be an arbitrarily chosen decomposing annulus. Then $S \cap A$ consists of curves that are essential in A and that are hence meridians of $\partial C(K)$. You will prove in the exercises that the complement of a non-trivial knot can't be a proper submanifold of a compression body. Hence S must meet each of the knot complements $C(K_i)$.

Now consider cutting S along all curves of intersection with the decomposing annuli. Let Q be a component of the resulting surface. We argue that Q can't be an annulus: Recall that the boundary components of Q are meridians of $\partial C(K)$. Hence if Q is an annulus, then it is either boundary parallel or it is a decomposing annulus. We assume the following: (1) that S intersects the collection of decomposing annuli in the fewest possible number of curves (subject to the condition that the curves of intersection are essential); (2) that each summand K_i is prime. (Otherwise we in fact have a connected sum of more than n knots.) Under these assumptions, if Q is an annulus, then it is parallel to one of the decomposing annuli. In particular, there must be some other component of $S \backslash \mathcal{S}$ in the knot complement containing Q.

It follows that each knot complement contains a non-annular component of $S \backslash \mathcal{S}$. Since S is orientable, a non-annular subsurface of S that has (non-empty) boundary consisting of curves essential in S can't be a sphere, disk, or torus. Hence it must have strictly negative Euler characteristic. In the exercises you will show that it also must have even Euler characteristic. Summing over the components of $C(K) \backslash \mathcal{S}$, this tells us that $\chi(S) \leq -2n$. Thus $g(S) \geq n+1$. □

How does this specialized result help us prove the theorem mentioned? It provides the key idea for a more formal proof. If our Heegaard splitting is weakly reducible, then we untelescope it to obtain a strongly irreducible generalized Heegaard splitting. The key idea helps guide us through a proof

6.9. An Application

of the theorem that relies on strongly irreducible generalized Heegaard splittings. In a generalized Heegaard splitting, we need to consider more than just one surface; we need to consider $(\bigcup_i F_i) \cup (\bigcup_i S_i)$. The following definition is crucial:

Definition 6.9.5. The *index* $J(V)$ of a compression body V is defined by the formula
$$J(V) - \chi(\partial_- V) - \chi(\partial_+ V).$$

Lemma 6.9.6. *If*
$$M = (V_1 \cup_{S_1} W_1) \cup_{F_1} (V_2 \cup_{S_2} W_2) \cup_{F_2} \cdots \cup_{F_{n-1}} (V_n \cup_{S_n} W_n)$$
is a strongly irreducible generalized Heegaard splitting that is an untelescoping of the Heegaard splitting
$$M = V \cup_S W,$$
then
$$-\chi(S) = -\chi(\partial_- V_1) + \chi \sum_i J(V_i).$$

You will prove Lemma 6.9.6 in the exercises. Now consider the decomposing annulus as it winds its way through a strongly irreducible generalized Heegaard splitting. See Figure 6.37.

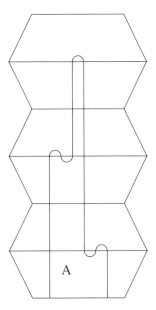

Figure 6.37. *Schematic for a decomposing annulus and a generalized Heegaard splitting.*

In the warm-up case considered above, we captured a portion of the splitting surface with negative Euler characteristic in each knot complement.

Now, we proceed similarly, but compute indices. We can do this by relying on the the following auxiliary lemma:

Lemma 6.9.7. *If $\mathcal{A}$ is a collection of annuli in a compression body V and X is a component of $V \setminus \mathcal{A}$, then*
$$\chi(\partial_- V \cap X) - \chi(\partial_+ V \cap X) \geq 0.$$
Furthermore, if
$$\chi(\partial_- V \cap X) - \chi(\partial_+ V \cap X) = 0,$$
then $V \cap X$ is a product.

You will prove Lemma 6.9.7 in the exercises. Consequently, each knot complement, since it is not a product, captures positive index. A counting argument then establishes the desired inequality genus$(S) \geq n + 1$. This in turn establishes the stated bound on the tunnel number.

Exercises

Exercise 1. Prove that a separating surface (possibly with boundary) in a knot complement has even Euler characteristic.

Exercise 2. Prove that the complement of a non-trivial knot can't be a proper submanfold of a compression body.

Exercise 3. Prove Lemma 6.9.6.

Exercise 4. Prove Lemma 6.9.7.

6.10. Heegaard Genus and Rank of Fundamental Group*

The description of a 3-manifold via a Heegaard splitting gives a natural way of computing the fundamental group of a 3-manifold. In this section we consider two distinct notions, the Heegaard genus of a 3-manifold and the rank of the fundamental group of a 3-manifold. The insight here translates into an inequality for these invariants.

Definition 6.10.1. The *Heegaard genus* of a 3-manifold M, denoted by $g(M)$, is the least possible genus of a splitting surface of a Heegaard splitting for M.

E.g., $g(\mathbb{S}^3) = 0$, $g(\text{lens space}) = 1$, $g(\text{prism manifold}) = 2$.

Definition 6.10.2. The *rank* of a 3-manifold M, denoted by $r(M)$, is the least number of generators required for $\pi_1(M)$.

Theorem 6.10.3. $r(M) \leq g(M)$.

6.10. Heegaard Genus and Rank of Fundamental Group

Proof. Given a Heegaard splitting $M = V \cup_S W$ that realizes $g(M)$, we may compute the fundamental group of M as follows: We consider M to be built from V in $g(M) + 1$ steps. At each of the first $g(M)$ steps, we add an open neighborhood of a 2-handle that can also be thought of as an open neighborhood of a meridian disk for W. In the final step, we add an open neighborhood of the 3-handle.

This description translates into a computation of $\pi_1(M)$. Here $\pi_1(V)$ is the free group on $g(M)$ generators. Adding an open neighborhood of a disk (whose fundamental group is trivial) adds a relation. Hence we add a relation at each of the first $g(M)$ steps. In the final step, a 3-ball (also with trivial fundamental group) is added along its boundary 2-sphere. Thus the fundamental group is unchanged.

We obtain a ("balanced") presentation
$$\pi_1(M) = \langle x_1, \ldots, x_{g(M)} \mid r_1, \ldots, r_{g(M)} \rangle. \qquad \square$$

The converse is not true; i.e., there are 3-manifolds for which the inequality is strict. See [**12**].

Theorem 6.10.4 (Schultens-Weidmann)**.** *Given any $n \in \mathbb{N}$, there is a 3-manifold M_n such that $g(M_n) - r(M_n) \geq n$.*

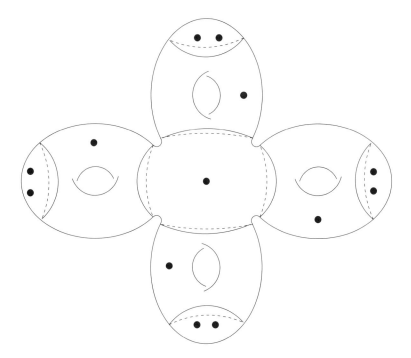

Figure 6.38. *A base orbifold considered in the proof of Theorem 6.10.4 (case $n = 4$).*

For a proof of Theorem 6.10.4, see [**144**]. The examples constructed in the proof are graph manifolds. A graph manifold is a 3-manifold modeled on a graph such that each vertex corresponds to a Seifert fibered space and each edge corresponds to the identification of two boundary components of these Seifert fibered spaces. The examples in question are modeled on star shaped graphs with $2n + 3$ vertices and n edges. In these examples, $\frac{r(M)}{g(M)}$ is roughly $\frac{3}{4}$. The minimum possible value for this expression is unknown. For more on this topic, see Li's treatment in [**86**].

Rank and genus are examples of "classical" invariants of 3-manifolds. Recent years have seen rapid progress in understanding "modern" invariants of 3-manifolds, growing out of homology theories. These topics go beyond the scope of this book. The reader is invited to peruse [**66**] and [**115**].

Exercises

Exercise 1. Assume that $\frac{r(M)}{g(M)} \geq \frac{3}{4}$ for all 3-manifolds M and prove that $r(M) = 0$ implies $M = \mathbb{S}^3$.

Exercise 2. Design a sufficiently complicated genus 2 Heegaard splitting and calculate the fundamental group of the 3-manifold using that Heegaard splitting.

Chapter 7

Further Topics

7.1. Basic Hyperbolic Geometry

Hyperbolic n-space can be realized in a variety of ways. We here discuss the upper half-space model. See also [8], [25], [74], [75], and [124]. Consider the set

$$\mathbb{U}^n = \{(x_1, \ldots, x_n) \in \mathbb{R}^n : x_n > 0\}.$$

The *element of hyperbolic arc length* on $\mathbb{U}^n$ is

$$ds = \frac{\sqrt{dx_1^2 + \cdots + dx_n^2}}{x_n}$$

and the *element of hyperbolic volume* is given by

$$dV = \frac{dx_1 \cdots dx_n}{x_n^n}.$$

This tells us how to calculate arc length and volume in hyperbolic space. See the 2-dimensional computations below.

Example 7.1.1. To calculate the arc length L of the horizontal path from $(0,1)$ to $(1,1)$ we parameterize the path by $x_1(t) = t$, $x_2(t) = 1$ for $t \in [0,1]$. Then,

$$L = \int_0^1 ds = \int_0^1 \frac{\sqrt{dx_1^2 + dx_2^2}}{x_2} = \int_0^1 \frac{\sqrt{1^2 + 0}}{1} dt = 1.$$

Example 7.1.2. For $n = 2$, to calculate the arc length L of the vertical path from $(0, a)$ to $(0, b)$, for $b > a > 0$, we parameterize the path by $x_1(t) = 0$,

$x_2(t) = t$. Then,

$$L = \int_a^b ds = \int_a^b \frac{\sqrt{dx_1^2 + dx_2^2}}{x_2}$$

$$= \int_a^b \frac{\sqrt{0+1^2}}{t} dt = [\ln t]_a^b$$

$$= \ln b - \ln a = \ln \frac{b}{a}.$$

In Exercise 1 of this section, you will show that a vertical arc minimizes the length among all arcs between $(0, a)$ and $(0, b)$.

Example 7.1.3. The subarc of the unit circle from $(0, 1)$ to $(1, 0)$ not including this second endpoint has infinite length. To see this, we parameterize this arc as $x_1(t) = \sin t$, $x_2(t) = \cos t$ and compute the following improper integral:

$$\int_0^1 ds = \int_0^{\frac{\pi}{2}} \frac{\sqrt{dx_1^2 + dx_2^2}}{x_2}$$

$$= \int_0^{\frac{\pi}{2}} \frac{\sqrt{\cos^2 t + \sin^2 t}}{\cos t}$$

$$= \int_0^{\frac{\pi}{2}} \frac{1}{\cos t} dt$$

$$= \lim_{a \to \frac{\pi}{2}} [\ln(\sec t + \tan t)]_0^a = \infty.$$

Example 7.1.4. To calculate the area of the region A between the two vertical rays given by $x_1 = \pm 1$ and above the semicircle $\{\ \mathbf{x}\ :\ |\mathbf{x}| = 1\ \}$ (see Figure 7.1), we compute the following integral:

$$\text{Area} = \int_A dV = \int_A \frac{dx_1 dx_2}{x_2^2}$$

$$= \int_{-1}^1 \int_{\sqrt{1-x_1^2}}^{\infty} \frac{dx_2 dx_1}{x_2^2}$$

$$= \int_{-1}^1 \left[-\frac{1}{x_2}\right]_{\sqrt{1-x_1^2}}^{\infty} dx_1$$

$$= \int_{-1}^1 \frac{1}{\sqrt{1-x_1^2}} dx_1$$

$$= [\arcsin x_1]_{-1}^1 = \pi.$$

Example 7.1.5. To calculate the area of the region A in the half-plane $x_1 \geq 0$, above the semicircle $\{\ \mathbf{x}\ :\ |\mathbf{x}| = 1\ \}$ and below the semicircle

7.1. Basic Hyperbolic Geometry

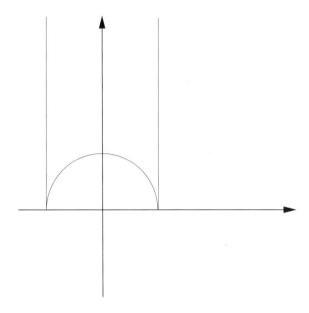

Figure 7.1. *A region with area π.*

$\{\,\mathbf{x} : |\mathbf{x} - 1| = 2\,\}$ (see Figure 7.2), we compute the following integral:

$$\text{Area} = \int_A dV = \int_A \frac{dx_1 dx_2}{x_2^2} = \int_0^1 \int_{\sqrt{1-x_1^2}}^{\sqrt{4-(x_1-1)^2}} \frac{dx_2 dx_1}{x_2^2}$$

$$= \int_0^1 \left[-\frac{1}{x_2}\right]_{\sqrt{1-x_1^2}}^{\sqrt{4-(x_1-1)^2}} dx_1$$

$$= \int_0^1 \left(-\frac{1}{\sqrt{4-(x_1-1)^2}} + \frac{1}{\sqrt{1-x_1^2}}\right) dx_1$$

$$= -\left[\arcsin\left(\frac{x_1-1}{2}\right)\right]_0^1 + [\arcsin x_1]_0^1$$

$$= -\frac{\pi}{6} + \frac{\pi}{2} = \frac{\pi}{3}.$$

In what follows, arc length will always be computed as above using the element of hyperbolic arc length. To summarize:

Definition 7.1.6. The *upper half-space model* for hyperbolic space is a metric space obtained as follows: Let $\mathbb{U}^n = \{(x_1, \ldots, x_n) \in \mathbb{R}^n \mid x_n > 0\}$. The distance between two points is the minimal length of an arc connecting the points. We denote this metric space by $(\mathbb{U}^n, d_{\mathbb{U}^n})$ or simply by $\mathbb{U}^n$.

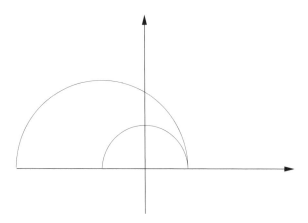

Figure 7.2. A region with area $\frac{\pi}{3}$.

Definition 7.1.7. Given a metric space (X, d), a *geodesic* in (X, d) is a path γ that is locally distance minimizing. I.e., for each $x \in X$, there is a neighborhood U such that for any two points in $\gamma \cap U$ the distance between these two points is the length of the subarc of $\gamma \cap U$ connecting them.

Definition 7.1.8. A *complete* geodesic is a bi-infinite geodesic.

Theorem 7.1.9. *The complete geodesics in $\mathbb{U}^n$ are vertical rays and half-circles that are orthogonal to $\mathbb{R}^{n-1} \times \{0\} \subset \mathbb{R}^n$. (The geodesics in $\mathbb{U}^n$ are subsegments of the complete geodesics.)*

For a proof of this fact see for instance John Radcliffe's book *Foundations of Hyperbolic Manifolds*, [**124**]. In attempting to understand a metric space, the symmetries of the space are of particular interest. Recall the notion of isometry from Chapter 1. It generalizes the notion of symmetry.

Example 7.1.10. Self-homeomorphisms of $\mathbb{U}^n$ of the form
$$(x_1, \ldots, x_n) \to (ax_1, \ldots, ax_n)$$
are called *dilations*. Since
$$ds = \frac{\sqrt{(dx_1)^2 + \cdots + (dx_n)^2}}{x_n}$$
$$\to \frac{\sqrt{(dax_1)^2 + \cdots + (dax_n)^2}}{ax_n} = \frac{\sqrt{a^2(dx_1)^2 + \cdots + a^2(dx_n)^2}}{ax_n} = ds$$
under this self-homeomorphism, dilations are isometries.

Example 7.1.11. Self-homeomorphisms of $\mathbb{U}^n$ of the form
$$(x_1, \ldots, x_n) \to (x_1 + b_1, \ldots, x_n + b_n)$$

7.1. Basic Hyperbolic Geometry

are called *translations*. Since

$$ds = \frac{\sqrt{dx_1^2 + \cdots + dx_n^2}}{x_n}$$

$$\rightarrow \frac{\sqrt{d(x_1 + b_1)^2 + \cdots + d(x_n + b_n)^2}}{x_n} = \frac{\sqrt{dx_1^2 + \cdots + dx_n^2}}{x_n} = ds$$

under this self-homeomorphism, translations are isometries.

Example 7.1.12. The self-homeomorphism of $\mathbb{U}^n$ given by

$$(x_1, \ldots, x_n) \rightarrow \left(\frac{x_1}{x_1^2 + \cdots + x_n^2}, \ldots, \frac{x_n}{x_1^2 + \cdots + x_n^2} \right)$$

is an example of an *inversion*. In the exercises, you will prove that this inversion is an isometry.

Example 7.1.13. The self-homeomorphisms of $\mathbb{U}^n$ of the form

$$\begin{bmatrix} A & 0 \\ 0 & 1 \end{bmatrix}$$

for $A \in SO(n-1)$ are called *rotations*. In the exercises, you will prove that rotations are isometries.

The group generated by translations acts transitively on vertical rays. The group generated by dilations, rotations, and translations acts transitively on half-circles orthogonal to $\mathbb{R}^{n-1} \times \{0\} \subset \mathbb{R}^n$. Also note that the inversion in the third example maps the vertical ray limiting on $(1/2, 0, \ldots, 0)$ to a half-circle orthogonal to $\mathbb{R}^{n-1} \times \{0\} \subset \mathbb{R}^n$ and limiting on $(0, \ldots, 0)$ and $(2, 0, \ldots, 0)$. Thus the group generated by dilations, translations, and the given inversion acts transitively on geodesics. In fact, you will show in the exercises that this group is the full group of isometries of $\mathbb{U}^n$. This group is denoted by either Isom($\mathbb{U}^n$) or Möb($\mathbb{U}^n$).

Definition 7.1.14. A *convex* subspace of $\mathbb{U}^n$ is a subset $X \subset \mathbb{U}^n$ such that X contains all geodesic segments between pairs of points in X.

Remark 7.1.15. The space $S^{n-1} = (\mathbb{R}^{n-1} \times \{0\}) \cup \{\infty\}$ is called the *sphere at infinity*. Thus, for instance, the top "end" of a vertical ray is said to lie on the sphere at ∞. Likewise, the two "ends" of a complete geodesic lie (at two distinct points) on the sphere at ∞.

Definition 7.1.16. A *hyperbolic triangle* is a 2-dimensional convex subset of $\mathbb{U}^n$ that is bounded by three geodesics (either geodesic segments or complete geodesics) that connect three points each of which lies either in $\mathbb{U}^n$ or on the sphere at ∞. If the hyperbolic triangle is bounded by three complete geodesics (and all "vertices" lie on the sphere at infinity), then the triangle is called an *ideal triangle*.

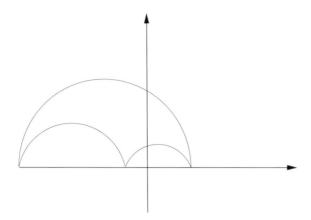

Figure 7.3. *An ideal triangle.*

In Example 7.1.4 we computed the area of a particular ideal triangle. It is π. In the exercises, you will show that all ideal triangles are isometric. A consequence of this is that the area of any ideal triangle is π.

Exercises

Exercise 1. Show that the distance in $\mathbb{U}^2$ between $(0, a)$ and $(0, b)$ is $\ln \frac{b}{a}$.

Exercise 2. Show that the inversion in Example 7.1.12 is an isometry.

Exercise 3. Show that rotations are isometries.

Exercise 4. Show that dilations, along with translations, as well as the one inversion given in Example 7.1.12 generate the full group of isometries of hyperbolic space, $\text{Isom}(\mathbb{U}^n)$.

Exercise 5. Use compositions of dilations, translations, and the inversion given in Example 7.1.12 to show that all ideal triangles are isometric.

7.2. Hyperbolic n-Manifolds**

In Chapter 1, we discussed examples of manifolds with additional structure. Hyperbolic manifolds are another such example. For more on hyperbolic manifolds, see [8], [**74**].

7.2. Hyperbolic n-Manifolds

Definition 7.2.1. A *hyperbolic n-manifold* is a manifold M with an atlas $\{(M_\alpha, \phi_\alpha)\}$ that satisfies the following additional requirements:
- for each α, $\phi_\alpha : M_\alpha \to \mathbb{U}^n$;
- for each pair α, β, the transition map $\phi_\beta \circ \phi_\alpha^{-1}$ is the restriction of an isometry of $\mathbb{U}^n$.

Example 7.2.2. Open subsets of $\mathbb{U}^n$ are hyperbolic n-manifolds.

A hyperbolic manifold M inherits the pull-back metric from its charts. The fact that transition maps are isometries guarantees that the metric is well-defined. Thus metric notions such as distances and geodesics are defined locally on M. Moreover, *angles* between two geodesics intersecting in a point can be measured by measuring the Euclidean angle between their tangents at the point.

From now on, unless stated otherwise, we will always assume that hyperbolic manifolds are connected and complete. If M is a compact manifold, then we say that M is hyperbolic if its interior is a hyperbolic manifold.

Below, we will be interested in a triangle with angles $(\frac{\pi}{4}, \frac{\pi}{8}, \frac{\pi}{8})$. To exhibit such a triangle, proceed as follows: First consider the ideal triangle bounded by the upper half of the unit circle and the vertical rays limiting on $(-1, 0)$ and $(1, 0)$ as in Figure 7.1.

The angles in this triangle are all 0. Now replace the vertical rays by large semicircles, orthogonal to $\mathbb{R}^{n-1} \times \{0\}$ that limit on $(-1, 0)$ and $(1, 0)$ and pass through $(0, t)$, for $t > 1$. See Figure 7.4. Then note that t can be chosen so that the angle at $(0, t)$ is $\frac{\pi}{4}$.

Next, we replace the upper half of the unit circle by a semicircle (centered at the origin) that passes through $(0, s)$, for $1 < s < t$. See Figure 7.5. As s increases continuously, the equal angles in the isosceles triangle increase continuously from 0 to $3\pi/8$. (In the limit, when $s = t$, the triangle is an infinitesimally small Euclidean triangle.) Thus for some intermediate value of s, we obtain angles of $\pi/8$. This provides a hyperbolic triangle with angles $(\frac{\pi}{4}, \frac{\pi}{8}, \frac{\pi}{8})$.

A similar argument can be used to show that for any $\alpha, \beta, \gamma \geq 0$ such that $\alpha + \beta + \gamma < \pi$, there is a hyperbolic triangle with angles α, β, γ. See the exercises.

Before discussing other examples, we wish to state the relation between hyperbolic n-manifolds and subgroups of $\text{Isom}(\mathbb{U}^n)$. To do so, we must define the following notion:

Definition 7.2.3. Let Γ be a group of homeomorphisms acting on a topological space X. A group Γ acts *properly discontinuously* if for any compact subset $K \subset X$, $K \cap gK$ is non-empty for only finitely many $g \in G$. Furthermore,

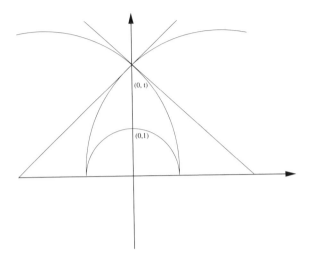

Figure 7.4. *Constructing the isosceles triangle.*

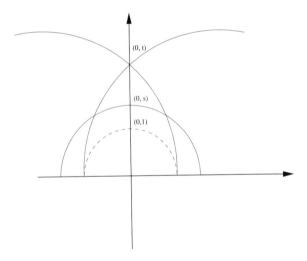

Figure 7.5. *Achieving the desired angles.*

Γ acts freely if for every $x \in X$, the stabilizer of x, $\Gamma_x = \{g \in \Gamma \mid gx = x\}$, is trivial.

Theorem 7.2.4. *M is a closed hyperbolic n-manifold if and only if it is the quotient of $\mathbb{U}^n$ by a subgroup Γ of isometries of $\mathbb{U}^n$ that acts freely and properly discontinuously on $\mathbb{U}^n$.*

The proof of this theorem is not difficult if one accepts the (non-trivial) fact that a simply connected hyperbolic n-manifold must be isometric to $\mathbb{U}^n$.

7.2. Hyperbolic n-Manifolds

Example 7.2.5. The closed orientable surface of genus 2 can be given a structure of a hyperbolic manifold. To see this, glue eight hyperbolic triangles with angles $(\frac{\pi}{4}, \frac{\pi}{8}, \frac{\pi}{8})$ by isometries in such a way that the eight angles of $\frac{\pi}{4}$ match up. This yields an octagon with angles of $\frac{\pi}{4}$.

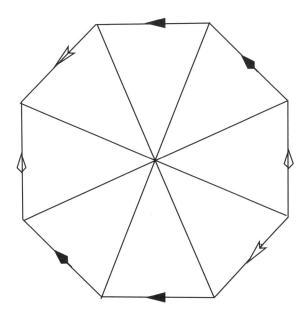

Figure 7.6. *An octagon with edge identifications.*

When we identify opposite edges in this octagon, as in Figure 7.6, we obtain a genus 2 surface with a hyperbolic structure.

Example 7.2.6. Higher genus closed orientable surfaces are also hyperbolic manifolds. Indeed, they can be realized as covering spaces of the genus 2 surface. The hyperbolic structure lifts to define a hyperbolic structure on the higher genus surface.

Example 7.2.7. The complement of the figure 8 knot.

In the exercises you will be invited to study Thurston's description of a hyperbolic structure for the figure 8 knot; see [**154**]. The study of low-dimensional hyperbolic manifolds was profoundly shaped by the work of William Thurston. In 1982 he was awarded the Fields Medal for his work. (He received the medal in 1983, due to the rescheduling of the ICM because of martial law in Poland.) Part of Thurston's vision was communicated in his book (see [**154**]) and lecture notes (see [**153**]). See also [**8**].

The contemplation of hyperbolic structures raises three questions: (1) Do hyperbolic structures exist on a given manifold? (2) If so, are they unique? (3) How do they help us to understand low-dimensional manifolds?

In partial answer to these questions, we saw in the previous section that closed orientable surfaces of genus at least 2 admit hyperbolic structures. Thus hyperbolic structures exist on these 2-dimensional manifolds. Note, however, that neither the sphere nor the torus admits a hyperbolic structure. In the case of 3-dimensional manifolds, theorems of Thurston and Perelman, stated below, provide necessary and sufficient conditions for 3-manifolds to possess a hyperbolic structure. Thus we have a complete understanding of the existence of hyperbolic structures in dimensions 2 and 3.

Definition 7.2.8. A 3-manifold M is *anannular* if it contains no essential annuli. It is *atoroidal* if every π_1-injective map of the 2-torus into M is homotopic into ∂M.

Theorem 7.2.9 (W. Thurston). *An anannular atoroidal Haken 3-manifold with (possibly empty) toroidal boundary admits a hyperbolic structure on its interior.*

The following theorem is known as the Hyperbolization Theorem. It was conjectured by Thurston and proved by Perelman. See [**121**], [**123**], and [**122**].

Theorem 7.2.10 (Perelman). *If M is a compact anannular atoroidal 3-manifold with (possibly empty) toroidal boundary whose universal cover is contractible, then M is hyperbolic.*

In Theorem 7.2.10, all the assumptions are necessary; this theorem does not hold for 3-manifolds in general. For instance, Seifert fibered spaces provide counterexamples if we drop the assumption of being atoroidal. The theorem below generalizes Theorem 7.2.10. It is known as the *Geometrization Conjecture* of Thurston.

Theorem 7.2.11 (Geometrization Conjecture). *Let M be a closed prime 3-manifold that does not contain 2-sided projective planes. Then M admits a decomposition along disjoint incompressible tori and Klein bottles into pieces M_j such that each M_j is geometric, i.e., each M_j is either hyperbolic or Seifert fibered.*

Perelman's outline of the proof of Theorem 7.2.11 can be found in [**121**], [**122**], [**123**]. For details, see [**9**], [**23**], [**27**], [**79**], [**103**], and [**104**]. For a larger context, see [**97**]. Perelman's work built on and realized Richard Hamilton's vision of the Ricci flow on manifolds; see [**54**], [**55**]. Perelman's methods and results provided profound insight into many issues pertaining to Riemannian 3-manifolds. In particular, he proved the Poincaré Conjecture. For this achievement, Perelman was awarded, but declined to accept, the Fields Medal in 2006. He was also awarded, but declined to accept,

7.2. Hyperbolic n-Manifolds

$1,000,000 awarded him by the Clay Institute for solving a Millennium Prize Problem, the Poincaré Conjecture.

In fact, hyperbolic structures not only exist on many manifolds but can be abundant. Closed orientable surfaces of genus ≥ 2, for instance, admit infinitely many distinct hyperbolic structures. The set of all hyperbolic structures on a given surface can be endowed with a suitable topology and is known as the Teichmüller space of the surface. In contrast, a (relatively) classical result in the study of hyperbolic manifolds is the following:

Theorem 7.2.12 (The Mostow Rigidity Theorem). *If M_1 and M_2 are homotopy equivalent hyperbolic n-manifolds (of finite volume) for $n \geq 3$, then M_1 and M_2 are isometric. In particular, they are homeomorphic. Furthermore, the homotopy equivalence is homotopic to an isometry.*

In the context of low-dimensional manifolds, we are interested in the topological implications of hyperbolic structures. Specifically, what does the fact that a 3-manifold possesses a hyperbolic structure tells us about the 3-manifold? In fact, current research often considers the space of all hyperbolic structures on a manifold. See, for instance, [17]. We now consider several applications of hyperbolic geometry to problems in topology.

We already saw an application of hyperbolic geometry in Section 2.6 when we alluded to an important theorem of Thurston. It deserves to be referenced again in this context: the classification of elements of the mapping class group into periodic, reducible, and pseudo-Anosov. We will discuss this classification in more detail in Section 7.6. The classification was inspired by work of Dehn and it was Nielsen who first used hyperbolic structures for its study. In the end, Thurston proved the Classification Theorem. In fact he provided more than one proof. One of his proofs employed hyperbolic geometry, specifically, the hyperbolic structures possessed by closed orientable surfaces of genus at least 2.

Another example of the use of hyperbolic geometry to solve problems in low-dimensional topology is the Smith Conjecture. It asks the following: Given a finite-order orientation-preserving diffeomorphism of the 3-sphere with non-empty fixed set, must the fixed set be the unknot? That this is indeed the case emerged from the work of several mathematicians, most notably Thurston, who employed hyperbolic structures on 3-manifolds. See [105].

Our final application of the use of hyperbolic geometry to study problems in low-dimensional topology is the more recent solution to the Tameness Conjecture. This line of investigation goes back to Waldhausen, who proved that the universal cover of a Haken 3-manifold is $\mathbb{R}^3$. See [156]. As a generalization of this result, Simon conjectured that if M is a compact 3-manifold and $\hat{M}$ is a covering space of M with finitely generated fundamental group, then $\hat{M}$ is tame. (Recall that a 3-manifold is *tame* if it is homeomorphic to the interior of a compact 3-manifold.) This conjecture was proved recently by a combination of the work of several people, most notably Perelman, Calegari-Gabai (see [20]), and Agol (see [3]). The crux of the matter was to handle the case of hyperbolic 3-manifolds. In this case it is not necessary to consider covering spaces. As proved by Agol and independently Calegari-Gabai, a hyperbolic 3-manifold with finitely generated fundamental group is tame. This fact was also known as the Marden Conjecture.

Exercises

Exercise 1. Let $\alpha, \beta, \gamma > 0$ with $\alpha + \beta + \gamma < \pi$. Construct a hyperbolic triangle with angles α, β, γ. (Hint: Proceed as in Example 7.2.2.)

Exercise 2. Show that a closed orientable surface of genus $n \geq 2$ is a hyperbolic manifold.

Exercise 3. Understand Thurston's example of a hyperbolic structure for the figure 8 knot complement. (This is described on pages 39 through 42 in his book.)

Exercise 4. Watch the movie *Not Knot*.

7.3. Dehn Surgery I

The idea behind Dehn surgery is simple: Given a knot (or link, respectively) in $K \subset \mathbb{S}^3$, set $C(K) = \mathbb{S}^3 \backslash \eta(K)$. Now create a new 3-manifold by attaching a solid torus (or solid tori, respectively) to the components of $\partial C(K)$. More needs to be said concerning the specifics of the regluing. The goal here is to obtain a new 3-manifold, not to simply reconstruct $\mathbb{S}^3$. For this reason we introduce a coordinate system. To do so, we must first prove a lemma.

Lemma 7.3.1. *Let $K \subset \mathbb{S}^3$ be a knot. Set $C(K) = \mathbb{S}^3 \backslash \eta(K)$. Let S_1, S_2 be Seifert surfaces for K. Then $S_1 \cap \partial C(K)$ is parallel to $S_2 \cap \partial C(K)$.*

Proof. Isotope S_1 and S_2 so that $S_1 \cap C(K)$ and $S_2 \cap C(K)$ intersect in arcs and simple closed curves. Note that if $S_i \cap C(K)$ is compressible, then we may perform compressions to obtain an incompressible surface. The result is a surface that is not homeomorphic to S_i but that has the same boundary as $S_i \cap C(K)$. Hence, as we are only interested in a statement about the boundaries of $S_1 \cap C(K)$ and $S_2 \cap C(K)$, we may assume that $S_1 \cap C(K)$ and $S_2 \cap C(K)$ are incompressible.

Now if there are simple closed curves in the intersection of $S_1 \cap C(K)$ and $S_2 \cap C(K)$, then they may be removed by a standard innermost disk argument. Further isotope $S_1 \cap C(K)$ and $S_2 \cap C(K)$ so that there are as few components of intersection as possible. Then consider an arc of intersection. Recall that Seifert surfaces are oriented. Along an arc of intersection, the right-hand rule induces an orientation on this arc. This orientation determines an initial + or − endpoint and a terminal − or + endpoint of the arc. In particular, the number of initial points is the same as the number of terminal points.

Now consider how the two torus knots $S_1 \cap \partial C(K)$ and $S_2 \cap \partial C(K)$ intersect on the torus $\partial C(K)$. Recall that they have been isotoped to intersect in as few points as possible. By marking the plus and minus sides of $S_1 \cap \partial C(K)$ and $S_2 \cap \partial C(K)$ on $\partial C(K)$, we can keep track of whether the points of intersection are initial or terminal points. Note, however, that as the torus knots intersect in a minimal number of points, they are either all initial points or all terminal points. Thus there can be no such intersections and the curves in $S_1 \cap \partial C(K)$ and $S_2 \cap \partial C(K)$ are parallel $\square$

Definition 7.3.2. Let $K \subset \mathbb{S}^3$ be a knot. Set $C(K) = \mathbb{S}^3 \setminus \eta(K)$. Denote by m the curve on $\partial C(K)$ that bounds a disk in $N(K)$. We call this curve the *meridian*. Any curve on $\partial C(K)$ that intersects m exactly once is called a *longitude*. Let S be a Seifert surface for K. Denote by l the curve $S \cap \partial C(K)$. We call this curve the *preferred* longitude.

The process of removing $\eta(K)$ from $\mathbb{S}^3$ and attaching a solid torus to the resulting 3-manifold in such a way that a meridian goes to a curve of slope m/l (i.e., wraps m times around the meridian and l times around the preferred longitude) on $\partial C(K)$ is called m/l-*Dehn surgery*. The case $r = 1/0$ is also called ∞-surgery or *trivial* surgery.

If M is a compact 3-manifold with a specified torus boundary component on which coordinates have been fixed, then the process of attaching a solid torus to M in such a way that the meridian goes to a curve of slope r on the specified torus boundary component of M is called a *Dehn filling* of M and is denoted by $M(r)$. Similarly, if M has several torus boundary components with specified coordinates, we write $M(r_1, \ldots, r_k)$ for the 3-manifold resulting from the appropriate sequence of Dehn fillings.

As an example, let us draw the preferred longitude of the trefoil. A Seifert surface of the trefoil is pictured in Figure 7.7.

Thus the preferred longitude is pictured in Figure 7.8.

Equivalently, the preferred longitude of the trefoil knot is pictured in Figure 7.9. In other contexts, the preferred longitude is called the *natural framing*. It is usually distinct from the so-called *blackboard framing* which derives its name from the fact that it is the closed curve that remains on the front side of a regular neighborhood of the knot as depicted on a blackboard.

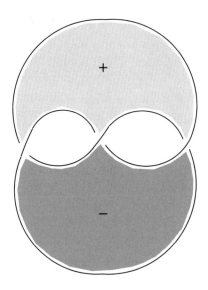

Figure 7.7. *The Seifert surface of the trefoil knot.*

Figure 7.8. *The preferred longitude of the trefoil knot.*

Being able to construct a given surface homeomorphism from a sequence of Dehn twists has profound ramifications in the study of 3-manifolds. As we will see, Heegaard splittings along with this description of surface homeomorphisms allows us to obtain a given 3-manifold via Dehn surgery on a knot or link.

7.3. Dehn Surgery I   229

Figure 7.9. *The preferred longitude of the trefoil knot.*

Lemma 7.3.3. *Let M be the connected sum of g factors of $\mathbb{S}^2 \times \mathbb{S}^1$. Then M is obtained by $(1/0,\ldots,1/0)$-Dehn surgery on the g component unlink in $\mathbb{S}^3$.*

Proof. We consider first the case in which $g = 1$. In this case $1/0$-Dehn surgery involves removing a regular neighborhood of the unlink, which creates a solid torus V, and then attaching a solid torus W to the resulting boundary component in such a way that a meridian of W goes to a meridian of V. This yields $\mathbb{S}^2 \times \mathbb{S}^1$.

More generally, consider the g component unlink. Separate the g components by a disjoint collection $\mathcal{S}$ of $g-1$ 2-spheres in $\mathbb{S}^3$. On each component of the unlink, perform $1/0$-Dehn surgery. Now $\mathcal{S}$ is a set of decomposing spheres that factors the resulting 3-manifold into g factors, each homeomorphic to $\mathbb{S}^2 \times \mathbb{S}^1$. □

Lemma 7.3.4. *Let S be a closed orientable surface of genus g. Let c be a simple closed curve in S. Let f be a Dehn twist around c. Let M_1 be the 3-manifold obtained by identifying two genus g handlebodies along their boundaries via f. Let M_2 be the 3-manifold obtained by identifying S with the splitting surface of the standard genus g Heegaard splitting of the connected sum of g factors of $\mathbb{S}^2 \times \mathbb{S}^1$ and then performing $1/1$-Dehn surgery along c. Then M_1 is homeomorphic to M_2.*

Proof. Both 3-manifolds in question have genus g Heegaard splittings. In the case of M_1, this follows from the construction. In the case of M_2,

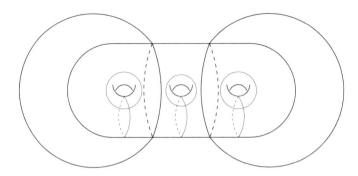

Figure 7.10. *Before the Dehn surgery.*

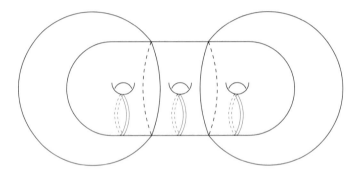

Figure 7.11. *After the Dehn surgery.*

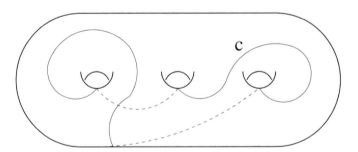

Figure 7.12. *The curve c.*

see Figure 7.11. In M_2, we can isotope c to lie just below the splitting surface and then consider the Dehn surgery to be taking place entirely in one handlebody. Denote this handlebody by H.

Let A be an annulus between c and the boundary of H. Denote a slightly truncated version of A lying in $H\backslash\eta(c)$ by A'. Now consider the effect of the Dehn surgery on c in H. It is the same as the effect of removing $\eta(c)$ from H, cutting along A', performing a full twist, regluing along A', and replacing $\eta(c)$.

Exercises

Figure 7.13. *Dehn surgery as Dehn twist.*

A curve on ∂H will have changed by a Dehn twist along c. In particular, the curves that specify how to attach the 2-handles of the complementary compression body will have changed by a Dehn twist along c. I.e., M_1 is homeomorphic to M_2. □

We now prove the main theorem:

Theorem 7.3.5. *Every closed orientable 3-manifold can be obtained by Dehn surgery on a link in $\mathbb{S}^3$.*

Proof. Let M be a closed orientable 3-manifold. Then M has a Heegaard splitting $M = V \cup_S W$. Let g be the genus of this Heegaard splitting. By Lemma 7.3.3 we obtain the connected sum of g copies of $\mathbb{S}^2 \times \mathbb{S}^1$ by Dehn surgery on a link in $\mathbb{S}^3$. Here the Heegaard splitting of the connected sum of g copies of $\mathbb{S}^2 \times \mathbb{S}^1$ has two copies of the handlebody V identified along their boundaries via the identity map.

By Theorem 2.6.5 we can factor f into Dehn twists, i.e., $f = f_1 \circ \cdots \circ f_n$, where each f_i is a Dehn twist around a curve c_i. Consider now the 3-manifold that is the connected sum of g factors of $\mathbb{S}^2 \times \mathbb{S}^1$. Further consider n parallel copies $S_1, \ldots, S_n$ of the splitting surface of this 3-manifold. Cutting along each surface S_i and reidentifying via the Dehn twist f_i yields M. By Lemma 7.3.4, this goal may also be attained via 1/1-Dehn surgeries along the curves c_i. □

Exercises

Exercise 1. Draw the longitude for the figure 8 knot.

Exercise 2. Generalize the notion of Dehn surgery on a knot to Dehn surgery on links.

Exercise 3. Describe how to obtain $L(2,1)$ by surgery on a link in $\mathbb{S}^3$.

Exercise 4. Describe how to obtain the 3-torus $\mathbb{S}^1 \times \mathbb{S}^1 \times \mathbb{S}^1$ by surgery on a link in $\mathbb{S}^3$.

7.4. Dehn Surgery II

In the 1980s Gordon and Luecke began approaching the study of 3-manifolds using Dehn surgery. Their school of thought caught on quickly and continues to produce results. See [46] and [30].

Definition 7.4.1. For M a compact 3-manifold with torus boundary and fixed coordinates on the torus boundary, the result of doing a Dehn filling along r is denoted by $M(r)$. If $M(r)$ has cyclic fundamental group, then the Dehn filling is called a *cyclic surgery*.

This definition is motivated by the example of knot complements. Knot complements have a single torus boundary component. As we saw above, the torus boundary component inherits a natural coordinate system. We are interested in those slopes for which Dehn surgery on the knot yields a 3-manifold with cyclic fundamental group.

Definition 7.4.2. Let r, s be torus knots that have been isotoped to intersect as few times as possible. Then we denote the number of intersections of r and s by $\Delta(r, s)$.

You will show in the exercises that $\Delta(m_1/l_1, m_2/l_2) = m_1 l_2 - m_2 l_1$.

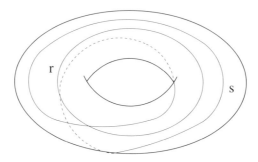

Figure 7.14. $\Delta(1/2, 0/1) = 1$.

Theorem 7.4.3 (The Cyclic Surgery Theorem by Culler-Gordon-Luecke-Shalen). *Suppose that M is not a Seifert fibered space. If $\pi_1(M(r))$ and $\pi_1(M(s))$ are cyclic, then $|\Delta(r, s)| \leq 1$.*

Theorem 7.4.4 (Gordon-Luecke). *If two knots have homeomorphic complements, then they are equivalent.*

One of the fundamental strategies employed by Gordon and Luecke was a combinatorial analysis of intersection patterns of surfaces. We will consider a few of the definitions involved and prove a lemma that illustrates the strength of these techniques.

7.4. Dehn Surgery II

Definition 7.4.5. Let M be a knot complement. Suppose P and Q are properly embedded compact connected planar incompressible surfaces in M that are isotoped to intersect in as few components as possible. Denote the slope represented by ∂P by π and the slope represented by ∂Q by γ.

Denote a capped off copy of P, i.e., a copy of P with a disk attached to each boundary component, by $\hat{P}$. Denote a capped off copy of Q by $\hat{Q}$. We construct graphs in $\hat{P}$ and $\hat{Q}$: Each disk that has been adjoined to P to form $\hat{P}$ is a "fat" vertex. Each arc of intersection of Q with P is an edge. This defines a graph in $\hat{P}$. We denote this graph by Γ_P. Analogously, we obtain a graph in $\hat{Q}$. We denote this graph by Γ_Q.

Remark 7.4.6. The minimality assumption on the number of components of intersection of P and Q guarantees that no arc of $P \cap Q$ is parallel to either ∂P or ∂Q.

Remark 7.4.7. There is a 1-to-1 correspondence between edges in Γ_P and Γ_Q.

Label the components of ∂P by $1, 2, \ldots, n_P$, likewise for ∂Q. Then it follows from our minimality assumptions on the number of components of intersection of $P \cap Q$ (and hence also $\partial P \cap \partial Q$) that on a boundary component of ∂Q we encounter intersections with components of ∂P labeled $1, 2, \ldots, n_P, 1, 2, \ldots, n_P, \ldots, 1, 2, \ldots, n_P$ and on a boundary component of ∂P we encounter intersections with components of ∂Q labeled $1, 2, \ldots, n_Q, 1, 2, \ldots, n_Q, \ldots, 1, 2, \ldots, n_Q$ in that order. See Figure 7.15.

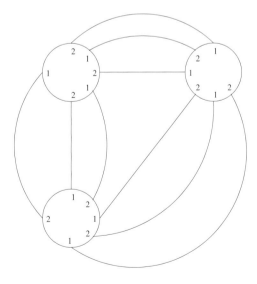

Figure 7.15. *Labels on Γ_P.*

Definition 7.4.8. An orientation of P (or Q, respectively) induces orientations on the components of ∂P. We say that two components of ∂P are *parallel* if they inherit the same orientation (on ∂M) and we say they are *antiparallel* if they inherit opposite orientations (on ∂M). We call vertices in Γ_P (or Γ_Q, respectively) *parallel* if they correspond to parallel components of ∂P (or ∂Q, respectively). We call vertices in Γ_P (or Γ_Q, respectively) *antiparallel* if they correspond to antiparallel components of ∂P (or ∂Q, respectively).

For an oriented edge, we denote the component of ∂e that has orientation -1 by $\partial_- e$, and the component that has orientation $+1$ by $\partial_+ e$. For an oriented vertex x, we denote the same vertex with the opposite orientation by $\bar{x}$.

The arcs of $P \cap Q$ are arcs of intersection of orientable surfaces in an orientable manifold. This leads to the following parity rule: If e is an edge of Γ_P that connects parallel vertices of Γ_P, then the labels at its endpoints represent antiparallel vertices of Γ_Q. If e is an edge of Γ_P that connects antiparallel vertices of Γ_P, then the labels at its endpoints represent parallel vertices of Γ_Q.

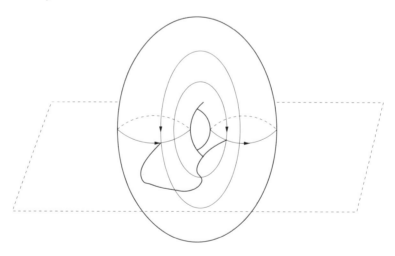

Figure 7.16. *Parity rule for an edge.*

Definition 7.4.9. An *x-cycle* in Γ_P or Γ_Q is a sequence of edges $e_1, \ldots, e_n$ such that the following hold for a suitable choice of orientations:

(1) the edges are all distinct;
(2) $\partial_- e_{i+1} = \partial_+ e_i$ for $i = 1, \ldots, n-1$ and $\partial_- e_1 = \partial_+ e_n$;
(3) the vertices ∂e_i are all parallel;
(4) the label at $\partial_- e_i$ is x for $i = 1, \ldots, n$.

7.4. Dehn Surgery II

Definition 7.4.10. A *Scharlemann cycle* is an x-cycle in Γ_P (or Γ_Q, respectively) such that the interior of one of the disks in $\hat{P}$ (or $\hat{Q}$, respectively) that the cycle bounds is disjoint from Γ_P (or Γ_Q, respectively).

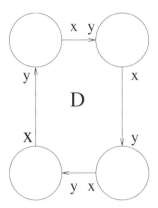

Figure 7.17. *A Scharlemann cycle.*

Definition 7.4.11. The *length* of a cycle is the number of edges it contains.

Scharlemann cycles play an important role in the combinatorics of intersection patterns of pairs of surfaces. Most importantly, they provide a strategy for simplifying such patterns. The following two lemmas give two immediate applications of Scharlemann cycles.

Lemma 7.4.12. *Neither Γ_P nor Γ_Q contains a Scharlemann cycle of length 1.*

Proof. (This is [30, Lemma 2.5.1].) Consider a cycle of length one in Γ_P. It corresponds to an arc of intersection of $P \cap Q$ with both ends on the same component of ∂P. If it is a Scharlemann cycle, then this arc of intersection cobounds, together with an arc in ∂M, a disk in P. Furthermore, the interior of this disk is disjoint from Γ_P and hence from Q. It therefore defines a boundary compression. This in turn defines a compression, but this is a contradiction because P was assumed to be incompressible. □

Lemma 7.4.13. *If Γ_P contains a Scharlemann cycle, then $M(\gamma)$ has a lens space summand.*

Proof. (This is [138, Proposition 5.6].) Denote by D the subdisk of $\hat{P}$ bounded by the Scharlemann cycle whose interior is disjoint from Γ_P. A posteriori, D lies in P since it is disjoint from Γ_P and hence does not meet the components of $\hat{P} \backslash P$ as these correspond to "fat" vertices of Γ_P. The boundary of D is partitioned into subarcs that lie alternately in ∂M and $P \cap Q$.

All edges in the Scharlemann cycle have the same label at their initial point. Recall the 1-to-1 correspondence between edges of Γ_P and Γ_Q. The fact that the edges of the Scharlemann cycle in Γ_P all have the same labels at their initial point means that they all correspond to edges in Γ_Q beginning at the same "fat" vertex of Γ_Q. Denote this "fat" vertex of Γ_Q by e_-.

Because the interior of D is disjoint from Γ_P, an edge in the Scharlemann cycle connects e_- to a vertex that corresponds to a component of ∂Q that is adjacent to the one corresponding to e_-. Recall that all vertices in a Scharlemann cycle are parallel. Recall also that if vertices are parallel in Γ_P, then they are antiparallel in Γ_Q. Thus an edge in the Scharlemann cycle in Γ_P corresponds to an edge in Γ_Q that connects e_- to an antiparallel vertex. There are at most two candidates for such a vertex, but since Q is orientable (so ∂D meets only one side of Q), there is exactly one such vertex. We denote it by e_+.

We can embed a copy of $\hat{Q}$ in $M(\gamma)$. Consider the solid torus J_γ that has been attached to M to obtain $M(\gamma)$. There are two meridian disks in J_γ bounded by the curves corresponding to e_- and e_+, respectively. We denote these by E_- and E_+, respectively.

Denote the 3-ball (solid cylinder) of ∂J_γ cobounded by ∂E_- and ∂E_+ that meets ∂D by B. Then $V = B \cup N(\hat{Q})$ is a solid torus. (Here $\hat{Q}$ is a 2-sphere and B is a 1-handle attached to $N(Q)$.) Thus $V \cup N(D)$ is a punctured lens space. □

Note, in particular, that the fundamental group of the punctured lens space obtained above is $\mathbb{Z}/p\mathbb{Z}$, where p is the number of edges in the Scharlemann cycle.

Definition 7.4.14. A knot K is said to satisfy *Property P* if any non-trivial Dehn surgery on K yields a 3-manifold that has non-trivial fundamental group.

In [11], Bleiler and Scharlemann show that this definition is equivalent to the following alternate definition:

Definition 7.4.15 (Alternate definition). A knot K is said to satisfy *Property P* if ± 1-surgery on K yields a 3-manifold that has non-trivial fundamental group.

Theorem 7.4.16. *Every non-trivial knot satisfies Property P.*

The proof of Theorem 7.4.16, due to Kronheimer and Mrowka, can be found in [82]. Earlier contributions can be found in [16], [34], [31], and

[41]. A related, but more specific, question is whether or not Dehn surgery on a knot in $\mathbb{S}^3$ can result in $\mathbb{S}^2 \times \mathbb{S}^1$.

Definition 7.4.17. A knot K is said to have *Property R* if no Dehn surgery on K yields $\mathbb{S}^2 \times \mathbb{S}^1$.

Somewhat more generally, one can ask about Dehn surgeries on knots that produce manifolds with an $\mathbb{S}^2 \times \mathbb{S}^1$ summand. Poenaru conjectured that no non-trivial such knots exist:

Conjecture 7.4.18 (Poenaru). *No surgery on a non-trivial knot yields a manifold with an $\mathbb{S}^2 \times \mathbb{S}^1$ summand.*

The following theorems were proved by Gabai; see [43], [44].

Theorem 7.4.19. *Every non-trivial knot satisfies Property R.*

Theorem 7.4.20. *The Poenaru Conjecture is true.*

The 2π-Theorem states conditions under which the result of Dehn filling a 3-manifold results in a hyperbolic 3-manifold. It was proved by Gromov and Thurston and improved by Agol and, from a different point of view, by Lackenby. We state it here in abbreviated form:

Theorem 7.4.21 (The 2π-Theorem). *Let M be a hyperbolic 3-manifold that is the complement of a knot in $\mathbb{S}^3$. With the exception of at most 12 slopes, the result of Dehn filling M is hyperbolic.*

More generally, one can ask what conditions allow the result of surgeries on a knot, or a class of knots, to have a given attribute, e.g., reducible, atoroidal, Haken, or hyperbolic. See the work of Rieck, Sedgwick, and others. Using Definition 7.4.22, graphic renditions of the results of such investigations have been given by Hall, Schleimer, and Segerman.

Definition 7.4.22. Let M be a compact 3-manifold with ∂M a torus. The set $\{M(r) \mid r \in \mathbb{Q}\}$ is called the *Dehn surgery space* of M.

Exercises

Exercise 1. Show that $\Delta(m_1/l_1, m_2/l_2) = m_1 l_2 - m_2 l_1$.

Exercise 2. What does the Cyclic Surgery Theorem tell us about the number of slopes such that Dehn surgery along that slope is a cyclic surgery?

Exercise 3. Describe how Properties P and R relate to the Poincaré Conjecture.

7.5. Foliations

Product manifolds enjoy a special place among manifolds due to their natural quotient maps. A question pertaining to a given product manifold often reduces to an analogous question on a manifold of lower dimension, where it is easier to solve. Of course, not every manifold is a product manifold. Foliations provide a local product structure that can be used in lieu of a global product structure. For example, recall how the product structure on $\mathbb{R}^3$ was used in our proof of the Schönflies Conjecture. Analogous arguments can be used on manifolds with appropriate foliations. For more on foliations, see [21], [22], [44], [50], [49], [51], and [114].

Definition 7.5.1. For $q \in [0, \infty]$, a C^q-codimension p *foliation* of an n-manifold M is a C^q-atlas with the following additional properties:

(1) for each chart (M_α, ϕ_α), M_α factors into $\mathbb{R}^p \times \mathbb{R}^{n-p}$;

(2) $\forall r \in \mathbb{R}^p \, \exists s \in \mathbb{R}^p$ such that the transition maps $\phi_{\alpha\beta} : M_\alpha \cap M_\beta \to M_\alpha \cap M_\beta$ satisfy $\phi_{\alpha\beta} : \{r\} \times \mathbb{R}^{n-p} \to \{s\} \times \mathbb{R}^{n-p}$.

We often denote a foliation by $\mathcal{F}$.

A maximal connected $(n-p)$-dimensional submanifold F such that each non-empty intersection $F \cap M_\alpha$ has the form (constants) $\times \mathbb{R}^{n-p}$ is called a *leaf* of the foliation.

Example 7.5.2. In Section 3.7 we discussed Seifert fibered spaces. Seifert fibrations constitute a special case of codimension 2 foliations of 3-manifolds; in fact, they provide all foliations of 3-manifolds whose leaves are circles. For a specific example, consider the foliation of $\mathbb{S}^3 = L(1,1)$ discussed there. This foliation is also known as the Hopf fibration of $\mathbb{S}^3$.

We will be particularly interested in codimension 1 foliations of surfaces and 3-manifolds. On a surface, a codimension 1 foliation decomposes the surface into leaves that are 1-manifolds. On a 3-manifold, a codimension 1 foliation decomposes the 3-manifold into leaves that are surfaces.

Example 7.5.3. The set of parallel lines of a fixed slope in $\mathbb{R}^2$ defines a codimension 1 foliation of $\mathbb{R}^2$. Likewise, the set of parallel planes in $\mathbb{R}^3$ defines a codimension 1 foliation of $\mathbb{R}^3$.

Example 7.5.4. The 2-torus $\mathbb{T}^2 = \mathbb{S}^1 \times \mathbb{S}^1$ admits a codimension 1 foliation with leaves $x \times \mathbb{S}^1$.

Example 7.5.5. The 3-torus $\mathbb{T}^3 = \mathbb{S}^1 \times \mathbb{S}^1 \times \mathbb{S}^1$ admits a codimension 1 foliation with leaves $x \times \mathbb{S}^1 \times \mathbb{S}^1$.

If a manifold M', covers another manifold, M, and M' has a foliation $\mathcal{F}'$ with charts that factor through the covering map, then $\mathcal{F}'$ projects to a foliation $\mathcal{F}$ of M. The leaves $\mathcal{F}'$ cover the leaves of $\mathcal{F}$. Examples 7.5.4 and 7.5.5 provide illustrations.

To construct many more examples, consider the covering $p : \mathbb{R}^n \to \mathbb{T}^n$ given by $p(x_1, \ldots, x_n) = (e^{i2\pi x_1}, \ldots, e^{i2\pi x_n})$ and foliations of $\mathbb{R}^n$ by parallel codimension k subspaces. Note that the leaves of the foliations induced on the n-torus can be compact or non-compact! E.g., in the case of $\mathbb{T}^2$, the leaves are circles if the slope of the lines in the cover are rational and they are lines if the slope of the lines in the cover are irrational.

Example 7.5.6. The infinite strip $\mathbb{R} \times [-1, 1]$ admits a foliation as pictured in Figure 7.18. The map $p : \mathbb{R} \times [-1, 1] \to \mathbb{S}^1 \times [-1, 1]$ given by $p(x, y) = (e^{i2\pi x}, y)$ is a covering that defines a foliation of the annulus. The foliation on the annulus thus obtained is called the *Reeb* foliation of the annulus.

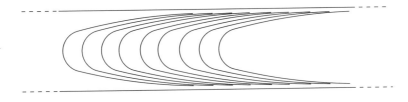

Figure 7.18. *A foliation of the infinite strip.*

Example 7.5.7. Rotating the above infinite strip around its core $\{0\} \times \mathbb{R}$ sweeps out an infinite solid cylinder and defines a foliation. The infinite solid cylinder covers the solid torus and defines a foliation of the solid torus. This foliation is called the *Reeb foliation* of the solid torus. See Figure 7.19.

Example 7.5.8. A codimension 1 foliation for the 3-sphere can be built from the genus 1 Heegaard splitting. The splitting surface is one leaf of the foliation and it separates the 3-sphere into two open solid tori foliated with Reeb foliations.

The only closed (connected) orientable surface to admit a codimension 1 foliation is the torus. (The technical reasons that preclude other closed orientable surfaces from having a codimension 1 foliation are that such a foliation induces a non-vanishing section of the projectivized tangent bundle of the surface S. This section lifts to a non-vanishing section of the tangent bundle of a 2-fold cover S' of S, i.e., a non-vanishing vector field on S. By the Poincaré-Hopf Index Theorem, the existence of such a vector field forces the Euler characteristic of S' to be 0. Since the Euler characteristic of a p-fold cover of S is $p \cdot \chi(S)$, we see that $\chi(S) = 0$; hence S is the torus.)

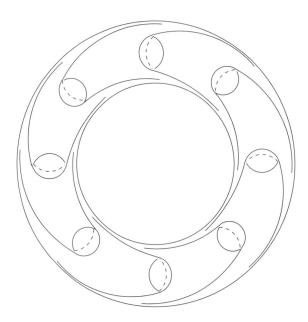

Figure 7.19. *The Reeb foliation of the solid torus.*

Recall that for a closed 3-manifold M, $\chi(M) = 0$ and there is no obvious obstruction to the existence of codimension 1 foliations. In fact, we have:

Theorem 7.5.9 (Lickorish-Zieschang). *Every closed orientable 3-manifold admits a codimension 1 foliation.*

Sketch of proof. Recall two of the theorems of Alexander: (1) *Every closed orientable 3-manifold is a branched cover of $\mathbb{S}^3$ over a link*; (2) *every link can be realized as a closed braid.* Also recall the foliation $\mathcal{F}$ of $\mathbb{S}^3$ given above.

Let M be a closed orientable 3-manifold and suppose that M is a branched cover of $\mathbb{S}^3$ over the link L. Isotope L to be a closed braid that lies in a small neighborhood of the core curve of one of the solid tori foliated by the Reeb foliation, so that each component L' of L intersects $\mathcal{F}$ transversely. We need only lift the foliation of $\mathbb{S}^3$ to a foliation of M. There is no obstruction to doing so: Away from L, the foliation lifts via a local homeomorphism; near L, it also lifts, in this case because L is transverse to $\mathcal{F}$. For more details, see [**44**]. □

The Reeb foliation plays an important part in the argument above for a good reason: There are 3-manifolds whose foliations necessarily contain a solid torus that is foliated by the Reeb foliation. However, for many applications, we are interested in precisely the foliations that contain no such solid torus.

7.5. Foliations

Definition 7.5.10. A foliation of a 3-manifold is *Reebless* if it contains no solid torus foliated by the Reeb foliation.

Definition 7.5.11. Let M be a 3-manifold with foliation $\mathcal{F}$. A *transversal* to $\mathcal{F}$ is a simple arc γ that meets $\mathcal{F}$ transversely. A *closed transversal* to $\mathcal{F}$ is a simple closed curve γ that meets $\mathcal{F}$ transversely.

The following theorem has been fundamental in the application of foliations to the study of 3-manifolds. It is a deep theorem and we will only sketch a proof of one of its parts.

Theorem 7.5.12 (S. Novikov [114])**.** *Let M be a closed orientable 3-manifold with a Reebless foliation $\mathcal{F}$. Then the following hold:*

(i) *if F is a leaf of $\mathcal{F}$, then F is incompressible in M;*

(ii) *if γ is a closed transversal to $\mathcal{F}$, then no power of γ is trivial in $\pi_1(M)$;*

(iii) *either $\pi_2(M) = 0$ or $\mathcal{F}$ is the product foliation on $M = \mathbb{S}^2 \times \mathbb{S}^1$;*

(iv) *M is not homeomorphic to S^3.*

Idea of proof of (i). Suppose that the simple closed curve c on the leaf F of $\mathcal{F}$ bounds a disk D in M but not in F. Consider the intersection of the leaves of $\mathcal{F}$ with D. See Figure 7.20.

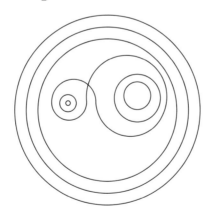

Figure 7.20. *Induced foliation on a disk.*

It is a non-trivial fact, necessitating the assumption that $\mathcal{F}$ be Reebless, that D can be isotoped so that away from a finite number of components of intersection, this intersection is a foliation $\mathcal{F}_D$ of D by closed curves. Moreover, the remaining components of intersection must be figure 8 curves and points.

If the number of such figure 8 curves is non-zero, we replace D by an innermost subdisk of D cut out by a figure 8 curve. Now D is a disk with a codimension 1 foliation away from a finite number of points. Recall the application of the Poincaré-Hopf Index Theorem discussed above. If D is a disk with a codimension 1 foliation away from a finite number of points, then the foliated subsurface of D must have Euler characteristic 0, hence must be an annulus. Specifically, D meets $\mathcal{F}$ in simple closed curves parallel to ∂D and one point. See Figure 7.21 for a cross-section of the disk relative to the foliation.

Locally, leaves of $\mathcal{F}$ are level sets of a Morse function f with no critical levels. Near the critical point of $f|_D$, say a maximum, D intersects leaves of $\mathcal{F}$ in inessential circles. Thus D can be isotoped downward until it lies in F.

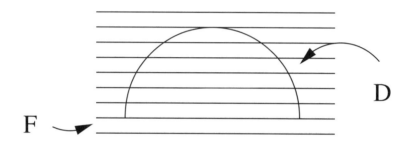

Figure 7.21. *Schematic for $\mathcal{F}$ near D.*

Arguing by induction on the number of figure 8 curves in the intersection of D with the leaves of $\mathcal{F}$ now provides an isotopy (relative to ∂D) of D into F, contradicting our initial assumption. □

Definition 7.5.13. A foliation $\mathcal{F}$ of M is *taut* if it has a closed transversal γ that intersects every leaf of $\mathcal{F}$.

Lemma 7.5.14. *A taut foliation must be Reebless.*

Proof. A closed transversal would not be able to enter and escape a solid torus foliated by the Reeb foliation. □

Novikov's Theorem has consequences:

Corollary 7.5.15. *If M is a closed orientable 3-manifold that admits a taut foliation, then $\pi_1(M)$ is infinite.*

Proof. By Lemma 7.5.14, the taut foliation, $\mathcal{F}$, is Reebless and admits a closed transversal. Thus M is a closed orientable 3-manifold that admits a Reebless foliation $\mathcal{F}$ and a closed transversal to $\mathcal{F}$; hence, by Novikov's Theorem, $\pi_1(M)$ is infinite. □

The above is just one of several useful properties enjoyed by 3-manifolds that admit taut foliations. The work of Candel, Lawson, Thurston, Sullivan, and others yields many more insights into foliations, especially their geometric structure. The work of Gabai provides specific applications of foliations to the study of knots and 3-manifolds.

Exercises

Exercise 1. Draw pictures of the foliations discussed in Examples 7.5.2 and 7.5.3. What is the closure of a leaf in Example 7.5.3?

Exercise 2. Prove that $\mathbb{T}^3$ admits infinitely many taut foliations.

Exercise 3. Prove that $\mathbb{S}^3$ does not admit a taut foliation.

Exercise 4. Describe several foliations of $\mathbb{T}^5$.

7.6. Laminations

We will focus our discussion of laminations mainly on the case of surfaces, that is, 2-dimensional manifolds, though the definitions apply in general. As we will see, the results for surfaces have direct applications in the world of 3-manifolds. For more detail, see [**25**], [**39**], [**45**], and [**120**].

Definition 7.6.1. A C^q-codimension p *lamination* of an n-manifold M is a C^q-codimension p foliation of a closed subset of M. The *leaves* of the lamination are the leaves of the foliation.

Example 7.6.2. A simple closed C^q-smooth curve in a surface S defines a C^q-codimension 1 lamination of S.

Example 7.6.3. Consider the annulus $\mathbb{A} = \mathbb{S}^1 \times I$ and let $C \subset I$ be the Cantor set. Then $\{\mathbb{S}^1 \times \{x\} \mid x \in C\}$ defines a codimension 1 lamination of $\mathbb{A}$.

Example 7.6.4. Figure 7.22 shows a codimension 1 lamination of the genus 2 surface. (Note how some leaves wind around others.)

Definition 7.6.5. Let S be a surface with a codimension 1 lamination Λ and let J be an arc in S that is transverse to Λ. A *transverse isotopy* of J is an isotopy of J such that J remains transverse to $\mathcal{F}$ and such that the endpoints of J either remain in the complement of Λ or remain in the same (respective) leaves during the entire isotopy.

A *transverse invariant measure* for Λ is a function f from arcs in S transverse to Λ into the set of measures on the arcs, $f(J) = \mu_J$, such that

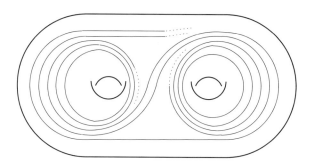

Figure 7.22. *A lamination on a genus 2 surface.*

the following hold:

- f is invariant under transverse isotopies of arcs;
- if $I \subset J$, then $\mu_I = \mu_J|_I$;
- $\mathrm{supp}(\mu_J) = J \cap \Lambda$.

A *measured lamination* on S is a lamination together with a transverse invariant measure.

The condition on the support of μ is not necessary, but it will simplify our discussion considerably. In particular, it guarantees that the topology defined below is Hausdorff.

Example 7.6.6. Let l be a simple closed curve in S and let $c \in \mathbb{R}^+$. Then l is a lamination. We can define a function f that takes arcs transverse to l and assigns them the number $c \cdot n$, where n is the number of times that the given arc intersects l. This turns l into a measured lamination called a *weighted curve*. Here c is called the *weight* of the simple closed curve.

Example 7.6.7. Let $\mathcal{F}$ be the foliation of $\mathbb{R}^2$ by parallel lines. We define a transverse measure to $\mathcal{F}$ by projecting onto a line orthogonal to the leaves of $\mathcal{F}$ and taking the length of the projection times some number $c > 0$.

Example 7.6.8. Consider the foliation of $\mathbb{T}^2$ given in Example 7.5.4. We can define a function f that takes arcs transverse to this foliation and assigns to a given arc J the number $c \cdot r$, where $c \in \mathbb{R}$ and r is the length of the arc obtained by projecting J onto the first factor in $\mathbb{S}^1 \times \mathbb{S}^1$.

There is a natural operation of multiplication of a measured lamination Λ by a positive real number k, denoted $k\Lambda$: The lamination itself does not change while the transverse measure is multiplied by k.

From now on, we will assume that S is a hyperbolic surface (with a fixed hyperbolic structure) and that our laminations are geodesic, that is, each

7.6. Laminations

leaf is a geodesic. The set of all measured (geodesic) laminations on S is denoted by $\mathcal{ML}(S)$.

A closed geodesic c on S provides a function $\hat{c} : \mathcal{ML}(S) \to \mathbb{R}$: For a given geodesic measured lamination Λ on S, either c is transverse to Λ or it is contained in a leaf of Λ. In the first case, $\hat{c}(\Lambda)$ is simply the measure of Λ evaluated on c, where we can think of c as partitioned into subarcs. In the second case, we decree that $\hat{c}(\Lambda) = 0$. We *topologize* $\mathcal{ML}(S)$ by giving it the weakest topology in which all functions $\hat{c}$ obtained from closed geodesics in this way are continuous. Thurston showed that weighted simple closed geodesics (as in Example 7.6.6) are dense in $\mathcal{ML}(S)$.

Definition 7.6.9. For weighted simple closed geodesics $(c, w), (c', w')$ in the closed hyperbolic surface S, the function i is defined by $i((c, w), (c', w')) = w \cdot w' \cdot n$, where n is the number of times that c intersects c' (transversely). The function i is called the *weighted intersection number*.

Theorem 7.6.10 (Penner)**.** *The function i extends continuously to $\mathcal{ML}(S)$.*

The extension of i to $\mathcal{ML}(S)$ is called the *intersection number* between measured laminations and is denoted

$$i : \mathcal{ML}(S) \times \mathcal{ML}(S) \to \mathbb{R}^+.$$

Theorem 7.6.11 (Thurston)**.** *Let S be a closed orientable surface of genus $g \geq 2$. There is a homeomorphism of $\mathcal{ML}(S)$ onto $\mathbb{R}^{6g-6}$. Moreover, this homeomorphism preserves multiplication by positive real numbers.*

Definition 7.6.12. We say that the measured laminations Λ and Λ' are *projectively equivalent* if they are equal as laminations and their transverse measures are positive scalar multiples of each other. The equivalence class $[\Lambda]$ of a (non-empty) measured lamination Λ under the equivalence relation defined by projective equivalence is called a *projective measured lamination*. The set of all projective measured laminations is denoted by $\mathcal{PML}(S)$.

Corollary 7.6.13. *Let S be a closed orientable surface of genus $g \geq 2$. Then $\mathcal{PML}(S)$ is homeomorphic to a sphere of dimension $6g - 7$.*

A closed hyperbolic surface S of genus $g > 1$ can be obtained from a $4g$-gon by identifying opposite sides. If the $4g$-gon is regular, then a rotation of the $4g$-gon through an angle of, say, $2\pi/4g$ induces a periodic automorphism of S of order $4g$.

Recall from Section 2.6 that every surface homeomorphism is a product of Dehn twists. A Dehn twist fixes the simple closed curve along which the Dehn twist is performed and is hence reducible. Likewise, a product of Dehn twists along a disjoint collection of simple closed curves is reducible.

Recall our brief discussion of the mapping class group in Section 7.3. Note that a mapping class (isotopy class of self-homeomorphisms of S) is periodic if it has a periodic representative and it is reducible if it has a reducible representative.

For the following definition, note that measured foliations are a special case of measured laminations.

Definition 7.6.14. Let S be a closed orientable surface and suppose $h : S \to S$ is a homeomorphism. If there are measured foliations $\mathcal{F}^+, \mathcal{F}^-$ of S such that $\mathcal{F}^+ \pitchfork \mathcal{F}^-$ and such that $h(\mathcal{F}^+) = k\mathcal{F}^+$ and $h(\mathcal{F}^-) = \frac{1}{k}\mathcal{F}^-$ for some $k > 1$, then h is called *Anosov*. A mapping class is called *Anosov* if it has an Anosov representative.

Example 7.6.15. Recall the covering of $\mathbb{T}^2$ by $\mathbb{R}^2$ discussed in Section 7.5. The matrix
$$A_h = \begin{bmatrix} 2 & 1 \\ 1 & 1 \end{bmatrix}$$
acts on $\mathbb{R}^2$ as a linear transformation and induces a self-homeomorphism h of $\mathbb{T}^2$ via the covering map $\mathbb{R}^2 \to \mathbb{T}^2$. The eigenvalues of A_h are
$$k = 3/2 + \sqrt{5}/2 \quad \text{and} \quad \frac{1}{k} = 3/2 - \sqrt{5}/2$$
and the corresponding eigenvectors are
$$(1, -1/2 + \sqrt{5}/2) \quad \text{and} \quad (1, -1/2 - \sqrt{5}/2).$$
We can foliate $\mathbb{R}^2$ by lines of slope $-1/2 + \sqrt{5}/2$ to obtain $\tilde{\mathcal{F}}^+$ and by lines of slope $-1/2 - \sqrt{5}/2$ to obtain $\tilde{\mathcal{F}}^-$. We endow these foliations with the transverse measures as in Example 7.6.7. Then $\tilde{\mathcal{F}}^\pm$ projects to a measured foliation $\mathcal{F}^\pm$ on $\mathbb{T}^2$. Moreover, $\mathcal{F}^+$ and $\mathcal{F}^-$ are transverse (in fact, their leaves are perpendicular). Finally, note that $h(\mathcal{F}^+) = k\mathcal{F}^+$ and $h(\mathcal{F}^-) = \frac{1}{k}\mathcal{F}^-$. Thus h is Anosov.

Due to the lack of foliations for closed oriented surfaces other than the torus, only maps of the torus have a chance of being Anosov. For this reason, the more general notion of pseudo-Anosov homeomorphisms given below is of interest.

Definition 7.6.16. A measured lamination Λ is *arational* if it contains no closed leaves. A lamination Λ is called *uniquely ergodic* if it admits a unique (up to scaling) transverse measure.

It is easy to see that every uniquely ergodic measure is arational (unless it consists of a single closed curve), while the converse is false.

Definition 7.6.17. A lamination Λ on a closed orientable surface S is said to be *filling* if $S \setminus \Lambda$ consists of disks.

7.6. Laminations

Definition 7.6.18. Let S be a closed orientable surface and suppose $h : S \to S$ is a homeomorphism. If there are filling measured laminations Λ^+, Λ^- of S such that $\Lambda^+ \pitchfork \Lambda^-$ and such that $h(\Lambda^+) = k\Lambda^+$ and $h(\Lambda^-) = \frac{1}{k}\Lambda^-$ for some $k > 1$, then h is called *pseudo-Anosov*. Here Λ^+ is called the *unstable* lamination and Λ^- is called the *stable* lamination.

A mapping class is called *pseudo-Anosov* if it has a pseudo-Anosov representative.

Theorem 7.6.19 (Thurston). *The stable and unstable laminations, $\Lambda^\pm$, of a pseudo-Anosov homeomorphism have the following properties:*

- $\Lambda^\pm$ *are arational;*
- $\Lambda^\pm$ *have no proper sublaminations; moreover, $\Lambda^\pm$ are uniquely ergodic;*
- $\forall [\Lambda'] \in \mathcal{PML}(S) \setminus \{\Lambda^-\}$ *the sequence $\{[h^n(\Lambda')]\}$ converges to $[\Lambda^+]$.*

Theorem 7.6.20 (Thurston). *Let S be a closed orientable surface and let $h : S \to S$ be an (orientation-preserving) homeomorphism. Then the mapping class $[h]$ is either periodic, reducible, or pseudo-Anosov.*

Proof in the case that S is a torus. First note that $[h]$ is determined by its action on $\pi_1(\mathbb{T})$. Moreover, this action is determined by a matrix, A_h, in $SL(2, \mathbb{Z})$. Understanding the structure of the matrix facilitates our understanding of the homeomorphism. Note that the characteristic polynomial for such a matrix is $t^2 - \text{trace}(A_h)t + 1$.

There are three cases: (1) $|\text{trace}(A_h)| < 2$; (2) $|\text{trace}(A_h)| = 2$; (3) $|\text{trace}(A_h)| > 2$. We consider each of these in turn.

(1) $|\text{trace}(A_h)| < 2$.

Since the entries in A_h are integers, the only possibilities are $\text{trace}(A_h) = 0, \pm 1$ (see Section 2.6). In each of these three subcases, we can compute the eigenvalues of A_h explicitly. They are $\pm i$ if $\text{trace}(A_h) = 0$, $1/2 \pm i(\sqrt{3}/2)$ if $\text{trace}(A_h) = 1$, and $-1/2 \pm i(\sqrt{3}/2)$ if $\text{trace}(A_h) = -1$. It follows that A_h is idempotent and hence that h is periodic.

(2) $|\text{trace}(A_h)| = 2$.

Since the entries in A_h are integers, the only possibilities are $\text{trace}(A_h) = \pm 2$. In both of these subcases, we can compute the eigenvalues of A_h explicitly; in fact, they are either both 1 or both -1. In the first case, A_h has an eigenvector corresponding to a curve on $\mathbb{T}^2$ that is fixed, so h is reducible. In the second case, A_h has an eigenvector corresponding to a curve on $\mathbb{T}^2$ that is fixed setwise, but reversed, so again h is reducible.

(3) $|\operatorname{trace}(A_h)| > 2$.

Here, too, A_h has two eigenvalues, but there are infinitely many possibilities for $\operatorname{trace}(A_h)$, so we cannot calculate them explicitly. Consideration of the characteristic polynomial of A_h (together with the fact that $\det(A_h) = 1$) tells us only that they are distinct real numbers λ, λ^{-1} with, say, $|\lambda| > 1 > |\lambda^{-1}|$. The eigenvector corresponding to λ determines a foliation of $\mathbb{T}$ as does the eigenvector corresponding to λ^{-1} (cf. Example 7.6.15). These two foliations satisfy the properties required for h to be an Anosov homeomorphism. □

Exercise

Exercise 1. Prove the following: All leaves of Λ are dense $\iff$ Λ has no proper sublaminations.

7.7. The Curve Complex

The curve complex, introduced by Harvey (see [57]), initially served to study topics related to surfaces and their automorphisms. In the late 1990s it reemerged, this time as a tool in the study of 3-manifolds, spurred on by a point of view proposed by J. Hempel; see [62]. We here provide a glimpse of those aspects of the curve complex most directly related to the topological study of 3-manifolds.

The curve complex exists on a surface. Through Heegaard diagrams and Heegaard splittings it allows us to describe and interpret characteristic features of 3-manifolds.

Definition 7.7.1. For any surface S, define the *complexity* of S, $c(S)$, by
$$c(S) = 3\operatorname{genus}(S) + \#|\partial S| - 4.$$
Let S be a surface with $c(S) > 0$. The curve complex of S is a simplicial complex with *simplices* as follows: The 0-simplices consist of isotopy classes of essential simple closed curves in S. The 1-simplices consist of pairs of 0-simplices that have disjoint representatives. More generally, the n-simplices consist of n-tuples of 0-simplices that have pairwise disjoint representatives. We denote the *curve complex* of S by $C(S)$.

The *distance* between two vertices of the curve complex is the minimal number of edges in an edge-path between the two vertices.

Note that for the excluded surfaces, the above definition would yield a highly disconnected complex. For this reason, one defines the curve complex slightly differently in these cases. For instance, on the torus no two non-isotopic essential curves are disjoint. There 1-simplices are defined to consist

7.7. The Curve Complex

of pairs of curves that have representatives intersecting exactly once. The resulting curve complex has the Farey graph as its 1-skeleton.

Definition 7.7.2. Let α, β be two simple closed curves in the surface S. The *geometric intersection number* of α and β, $I(\alpha, \beta)$, is the minimal number of times a curve isotopic to α intersects a curve isotopic to β.

Lemma 7.7.3. *Let S be a surface with $c(S) > 0$. For vertices x, y of $C(S)$ and representatives α_x, α_y of x, y, respectively, with $I(\alpha_x, \alpha_y) > 0$, we have*

$$d(x, y) \leq 2 + 2\log_2(I(\alpha_x, \alpha_y)).$$

Proof. Our proof proceeds by induction on $I(\alpha_x, \alpha_y)$. If $I(\alpha_x, \alpha_y) = 1$, then α_x and α_y can be isotoped to intersect once. Consider a collar of $\alpha_x \cup \alpha_y$, $C(\alpha_x \cup \alpha_y)$. Here $\partial C(\alpha_x \cup \alpha_y)$ is a curve that is disjoint from both α_x and α_y. This curve does not bound a disk in $C(\alpha_x \cup \alpha_y)$ and also does not bound a disk in the complement of $C(\alpha_x \cup \alpha_y)$ (indeed, this would force S to be a torus). Thus $d(x, y) = 2$. Since $2 \leq 2 + 0$, the inequality holds.

Now suppose that $I(\alpha_x, \alpha_y) \geq 2$. (Here it does not help to consider the collar of $\alpha_x \cup \alpha_y$, because its boundary consists of several curves, all of which could be inessential.) Consider a subarc a of α_x that has its endpoints on α_y and interior disjoint from α_y. We may assume that α_x and α_y have been isotoped to intersect in a minimal number of points. It then follows that a is not parallel to α_y.

The endpoints of a cut α_y into two subarcs. Let b be the subarc that meets α_x no more times than the other (i.e., fewer, if possible). Now a push-off of $a \cup b$, which we denote by c (see Figure 7.23) and whose isotopy class we denote by z, intersects α_y at most once and α_x at most $\frac{1}{2}I(\alpha_x, \alpha_y)$ times. Thus, by induction,

$$d(x, y) \leq d(x, z) + d(z, y) \leq 2 + (2 + 2\log_2(I(c, \alpha_y)))$$
$$\leq 4 + 2\log_2\left(\frac{1}{2}I(\alpha_x, \alpha_y)\right) = 4 + 2(-1 + \log_2 I(\alpha_x, \alpha_y))$$
$$= 2 + 2\log_2 I(\alpha_x, \alpha_y),$$

as required. □

Theorem 7.7.4. *For a surface S with $\chi(S) \leq -2$, the curve complex, $C(S)$, is connected.*

Proof. This follows immediately from Lemma 7.7.3. □

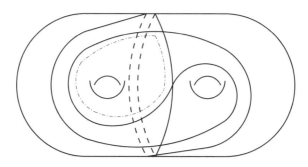

Figure 7.23. *The curves α_x, α_y, and c.*

We list two poignant results that provide structural information concerning the curve complex. The second requires a few definitions:

Theorem 7.7.5 (Hempel). *For $h : S \to S$ a pseudo-Anosov automorphism,*
$$\lim_{n \to \infty} d(x, h^n(x)) = \infty.$$
Consequently, the diameter of $C(S)$ is infinite.

The proof of this theorem relies on measured laminations and their relation to pseudo-Anosov automorphisms. The idea of the proof is to argue by contradiction and show that boundedness of the set $\{d(x, h^n(x)) \mid n \in \mathbb{N}\}$ contradicts key properties concerning the behavior of laminations with respect to pseudo-Anosov automorphisms. We mention these key properties as needed.

Proof. Suppose that $\{d(x, h^n(x)) \mid n \in \mathbb{N}\}$ is bounded. By passing to a subsequence, if necessary, we may assume that
$$\lim_{n \to \infty} d(x, h^n(x)) = m$$
and
$$d(x, h^n(x)) = m \ \forall n.$$

This assumption guarantees that for each n, there is a sequence $y_0^n, \ldots, y_m^n$ (corresponding to a row in the array below) such that $y_0^n = x$, $y_m^n = h^n(x)$, and $d(y_j^n, y_{j+1}^n) = 0$. Thus for each j, we obtain a sequence $\{y_j^n \mid n\}$ (corresponding to columns in the array below):

$$\begin{array}{ccccc} x & y_1^1 & \cdots & y_{m-1}^1 & h(x) \\ x & y_1^2 & \cdots & y_{m-1}^2 & h^2(x) \\ x & y_1^3 & \cdots & y_{m-1}^3 & h^3(x) \\ \cdots & \cdots & \cdots & \cdots & \cdots \end{array}$$

Note that a curve is also a lamination. Recall Theorem 7.6.11 which tells us that $\mathcal{ML}(S)$ is homeomorphic to $\mathbb{R}^{6g-6}$. Thus by adding appropriate weights, we can ensure that all the laminations under consideration lie in the unit sphere of $\mathbb{R}^{6g-6}$:

$$\begin{array}{cccccc}
(x,w) & (y_1^1, w_1^1) & \cdots & (y_{m-1}^1, w_{m-1}^1) & (h(x), w^1) \\
(x,w) & (y_1^2, w_1^2) & \cdots & (y_{m-1}^2, w_{m-1}^2) & (h^2(x), w^2) \\
(x,w) & (y_1^3, w_1^3) & \cdots & (y_{m-1}^3, w_{m-1}^3) & (h^3(x), w^3) \\
\cdots & \cdots & \cdots & \cdots & \cdots
\end{array}$$

Since the unit sphere of $\mathbb{R}^{6g-6}$ is compact, we may assume, by passing to a subsequence, if necessary, that the *measured laminations* forming the columns in this array converge to limits. Denote the limiting lamination of the sequence $\{(y_j^n, w_j^n) \mid n\}$ by (L_j, w_j). The curves y_j^n, y_{j+1}^n are disjoint; i.e., $i((y_j^n, w_j^n), (y_{j+1}^n, w_{j+1}^n)) = 0$ and the function i is continuous. Hence $i((L_j, w_j), (L_{j+1}, w_{j+1})) = 0$.

The measured lamination obtained as a limit of iterated powers of h,

$$\lim_{n \to \infty} \{(h^n(x), w^n)\} = (\Lambda^-, w^\infty),$$

is the stable lamination of h. It has special properties. One key property is that for any projective measured lamination μ, either $\mu = \Lambda$ or $i(\mu, \Lambda) \neq 0$. Since $i((L_{m-1}, w_{m-1}), (\Lambda, w^\infty)) = 0$, we have

$$(L_{m-1}, w_{m-1}) = (\Lambda, w^\infty).$$

In particular, L_{m-1} has the same properties as Λ. This in turn implies that

$$(L_{m-2}, w_{m-2}) = (L_{m-1}, w_{m-1}) \cdots (L_1, w_1) = (x, w).$$

However, another key property is that it is not possible for the stable lamination of a pseudo-Anosov automorphism to be a weighted curve. Thus we obtain a contradiction. $\square$

H. Masur and Y. Minsky studied the curve complex extensively; see [94] and [95]. They used a variety of methods, most notably those from geometric group theory. One of their most striking results relies on the following definition, due to M. Gromov:

Definition 7.7.6. A triangle in a metric space is δ-*thin* if each of its sides lies within a δ-neighborhood of its other two sides. A metric space is called *geodesic* if every two points are connected by a distance-minimizing geodesic. A geodesic metric space is *Gromov hyperbolic* if for some $\delta < \infty$, every triangle is δ-thin.

Theorem 7.7.7 (Masur-Minsky). *If S is a surface with $c(S) > 0$, then $C(S)$ is Gromov hyperbolic.*

The proof of this theorem is lengthy and requires several techniques from geometric group theory and Teichmüller theory that go beyond the scope of this book. Bowditch has a more accessible proof. See [**15**].

An element of the mapping class group of a surface S takes essential curves to essential curves. Moreover, it takes disjoint essential curves to disjoint essential curves. Thus it induces a map on $C(S)$. In this way the mapping class group acts not only on surfaces but also on their associated curve complexes. The converse is partially true:

Theorem 7.7.8 (Luo). *Let S be a surface not equal to a twice-punctured torus and suppose that $3\operatorname{genus}(S) + \#|\partial S| - 4 \geq 1$. Then every automorphism of $C(S)$ is induced by an automorphism of S.*

For S the twice-punctured torus, there is an automorphism of $C(S)$ that is not induced by an automorphism of S.

For a proof of Theorem 7.7.8 see [**91**].

Exercises

Exercise 1. Draw a picture of a genus 2 surface and a pair of curves of distance 2.

Exercise 2. Prove that hyperbolic n-space is δ-thin for $\delta = 1$.

Exercise 3. Prove that the dimension of $C(S)$ is $3\operatorname{genus}(S) + \#|\partial S| - 4$.

7.8. Through the Looking Glass**

In this section we explore some of the ideas, results, and ramifications of Hempel's work in "3-manifolds viewed from the curve complex"; see [**62**] and [**15**]. Here the curve complex can be thought of as a looking glass that allows for a view of 3-manifolds complementary to the view from within the 3-manifold.

Definition 7.8.1. Let S be a surface and X a simplex of $C(S)$. We denote by V_X the result of attaching 2-handles to $S \times I$ along the curves $c \times 1$ for $c \in X$ and attaching 3-handles whenever possible. Denote by K_X the subcomplex of $C(X)$ consisting of all simplices Y such that $V_Y = V_X$.

The distance between subcomplexes of $C(S)$ is the minimum distance between vertices in the subcomplexes.

Given a Heegaard splitting $M = V \cup_S W$, denote the simplex in $C(S)$ spanned by a given set of meridian disks of V by X and the simplex spanned by a given set of meridian disks of W by Y. The *distance of the Heegaard splitting* $M = V \cup_S W$, $d(M = V \cup_S W)$, is the *distance*, in $C(S)$, between the subcomplexes K_X and K_Y.

7.8. Through the Looking Glass

In effect, the distance of a Heegaard splitting is the smallest possible distance between a meridian disk for one handlebody and a meridian disk for the other handlebody.

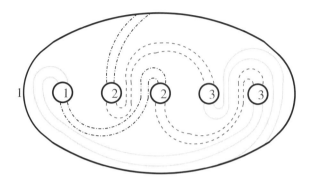

Figure 7.24. *A Heegaard diagram for the standard Heegaard splitting of $\mathbb{T}^3$.*

Example 7.8.2. Earlier, we discussed the standard Heegaard splitting of $\mathbb{T}^3$ and a diagram (see Figure 7.24) for this Heegaard splitting. Recall that this Heegaard splitting is weakly reducible. Specifically, note that, for instance, the disk in V corresponding to the disk labeled 2, that V was cut along, and the disk in W, corresponding to the solid curve, are disjoint. The former corresponds to a vertex in K_X, the latter to a vertex in K_Y. Thus $d(M = V \cup_S W) = 1$.

Consider what it means for a Heegaard splitting to have distance greater than 2. Suppose that a is a curve that bounds a disk in one handlebody and b is a curve that bounds a disk in the other handlebody. A distance greater than 2 means that these curves meet and that the complement of their union contains no essential curves. This makes the definition below relevant.

Definition 7.8.3. Two curves in a surface, isotoped to realize their intersection number, are said to *fill* the surface if their complement consists of disks.

The following conjectures and theorems show how the curve complex provides a lens through which to see and understand 3-manifolds.

Conjecture 7.8.4 (Schleimer). *A 3-manifold has only finitely many Heegaard splittings of distance greater than 2.*

Theorem 7.8.5 (Hempel). *A Heegaard splitting of a Seifert fibered space has distance at most 2.*

Proof. We prove this theorem in the special case of prism manifolds. As discussed in Section 6.5, these manifolds have only vertical and horizontal Heegaard splittings. Recall Definition 6.5.6. The cocore of a regular neighborhood of the arc γ is a disk D. The annulus in the prism manifold that projects to γ contains a disk E. These two disks intersect twice along their boundaries. Hence our conclusion follows from the argument in the proof of Lemma 7.7.3. (This argument holds in the more general setting of vertical Heegaard splittings for arbitrary Seifert fibered spaces.)

Recall also the informal discussion of horizontal Heegaard splittings. A horizontal Heegaard surface comes from two copies of F in the surface bundle $F \times I/\phi$ away from some exceptional fiber e. Because $F \times I/\phi$ is Seifert fibered, ϕ must be periodic.

Consider an arc a in F. It gives rise to two disks ($a \times [0, 1/2]$ and $a \times [1/2, 1]$) that extend to disks in the two handlebodies of the Heegaard splitting. The corresponding disks can be made disjoint in $F \times 1/2$ (where they are parallel) but will intersect along $F \times 0 = F \times 1$. Because ϕ is periodic, it does not matter that they intersect, because they do not fill F. This gives us a curve on F that is disjoint from both disks. □

Theorem 7.8.6 (Hartshorn). *A Heegaard splitting of a toroidal 3-manifold has distance at most 2.*

The proof of Theorem 7.8.6 can be found in [56]. It can be generalized to Heegaard splittings of 3-manifolds containing incompressible surfaces. The following idea is key:

Lemma 7.8.7. *A connected surface S in a compression body W with $\partial S \subset \partial_+ W$ is either a disk or it is compressible or it is boundary compressible.*

Proof. Denote the incompressible surface by S and the compression body by W. Choose a set $\mathcal{D}$ of meridian disks of W. We may assume that S intersects $\mathcal{D}$ in as few components as possible. A standard innermost disk argument then shows that all closed components of intersection define compressing disks. Furthermore, a standard outermost arc argument shows that all arcs of intersection define boundary compressions. Hence we need only consider the case in which S is disjoint from $\mathcal{D}$.

Here S lies in (closed orientable surface) $\times I$ union 3-balls. You will prove in the exercises that if it lies in (closed orientable surface) $\times I$, then it is either compressible or boundary parallel. In particular, it is either a disk or is compressible or is boundary compressible. If S lies in a 3-ball, then ∂S is a circle on the boundary of a 3-ball. Hence here, too, S is either a disk or is compressible. □

7.8. Through the Looking Glass

We can now outline a proof of Hartshorn's Theorem: Let M be a toroidal 3-manifold with Heegaard splitting $M = V \cup_S W$. Denote an essential torus in M by T. We may assume that the number of components of $S \cap T$ is minimal. Then any component $S \cap T$ that is inessential in T must be essential in S. An innermost such component provides a curve on S bounding a meridian disk in either V or W.

If there is no such component bounding a meridian disk in, say, V, then $T \cap V$ is boundary compressible. The boundary compression raises the Euler characteristic of $T \cap V$ by 1. Furthermore, all resulting curves of intersection between T and S can be made disjoint from the original curves of intersection and hence are distance 1 from all such curves. At this point a series of classical, but technical, results guarantees that under our minimality assumption, there must be a meridian disk for V in $T \cap V$ after the boundary compression.

Conjecture 7.8.8 (Hempel). *If the distance of a Heegaard splitting is sufficiently large, then the 3-manifold represented by the Heegaard splitting is hyperbolic.*

In his dissertation, Hossein Namazi considered this conjecture together with related questions. This led to a collaboration with Juan Souto. The duo proved several striking results, some of which we mention below. Many of these theorems can be interpreted as descriptions of "generic" behavior for Heegaard splittings.

Theorem 7.8.9, Theorem 7.8.11, and Corollary 7.8.12 use the notion of distance to interpolate between topology and geometry. They make extensive use of Riemannian geometry. Corollary 7.8.12, in addition, requires familiarity with algebraic topology.

Given a Heegaard splitting $M = V \cup_S W$, we consider $\mathcal{MCG}(S)$ of (an abstract copy of) S. We then consider the subgroup of $\mathcal{MCG}(S)$ that extends to both V and W (i.e., the intersection of the kernels of the maps $(i_V)_* : \mathcal{MCG}(S) \hookrightarrow \mathcal{MCG}(V)$ and $(i_W)_* : \mathcal{MCG}(S) \hookrightarrow \mathcal{MCG}(W)$ induced by the inclusion maps $i_V : S \hookrightarrow V$ and $i_W : S \hookrightarrow W$). The following theorem concerns this group.

Theorem 7.8.9 (Namazi). *If the distance of a Heegaard splitting $M = V \cup_S W$ is sufficiently large, then the subgroup of the mapping class group of S that extends to both V and W is finite.*

For a proof of Theorem 7.8.9, see [**111**]. In the preceding section we saw that V defines a subcomplex K_X of $\mathcal{C}(\partial_+ V)$. Note that $\mathcal{C}(\partial_+ V) \subset \mathcal{PML}(\partial_+ V)$. Consider the following setup: V_1, V_2 are compression bodies with $\partial_+ V_1$ and $\partial_+ V_2$ homeomorphic hyperbolic surfaces identified with the abstract surface S.

Definition 7.8.10. The map $h : \partial_+ V_1 \to \partial_+ V_2$ is a *generic* pseudo-Anosov if $h \in \mathcal{MCG}(S)$ is pseudo-Anosov, the stable lamination of h is not contained in $K_{V_1}^c$, and the unstable lamination of h is not contained in $K_{V_2}^c$.

Theorem 7.8.11 (Namazi-Souto). *Let V_1, V_2 be compression bodies with $\partial_+ V_1$ and $\partial_+ V_2$ homeomorphic hyperbolic surfaces identified with an abstract surface S and let $h : \partial_+ V_1 \to \partial_+ V_2$ be a generic pseudo-Anosov. For every $\epsilon > 0$, there exists n_ϵ such that if $n \geq n_\epsilon$, then $M_n = V_1 \cup_{h^n} V_2$ admits a Riemannian metric $\rho_{n,\epsilon}$ with all sectional curvatures pinched between $-1 - \epsilon$ and $-1 + \epsilon$. Moreover, $\rho_{n,\epsilon}$ has a lower bound for its injectivity radius independent of n and ϵ.*

Corollary 7.8.12. *Under the assumptions above, $\pi_1(M_n)$ is infinite and word hyperbolic.*

Perelman's work then implies that M is hyperbolic.

Considering sequences of manifolds raises questions concerning limiting behavior. There are several ways to define limits of such a sequence. One such way grows out of the notion of Hausdorff distance and was formulated by Gromov:

Definition 7.8.13. We say that the *Hausdorff distance between the Riemannian manifolds M, N is less than ϵ* if there is a diffeomorphism $q : M \to N$ such that

$$d_M(x,y) - \epsilon \leq d_N(q(x), q(y)) \leq d_M(x,y) + \epsilon.$$

A map such as q is called a $(1, \epsilon)$-*quasi-isometry*.

The notion of Hausdorff distance (and the related notion of Hausdorff topology) is well adapted to the study of compact spaces. For instance, it enables a discussion of convergent sequences. Gromov's generalization takes the notion a step further and thereby adapts it to the setting of non-compact spaces.

Definition 7.8.14. A sequence of pairs (M_n, p_n), where M_n is a manifold and p_n is a point in M_n, converges in the *Gromov-Hausdorff topology* if for every $R > 0$, the balls

$$B_n = \{x \in M_n \mid d_{M_n}(x, p_n) \leq R\}$$

converge in the Hausdorff topology.

For example, the sequence of circles in $\mathbb{R}^2$ with increasing radii and based at $(0,0)$ pictured in Figure 7.25 will converge to the x-axis in the Gromov-Hausdorff topology. For details on the results of Namazi and Souto, see [**112**].

7.8. Through the Looking Glass

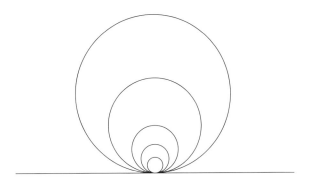

Figure 7.25. *A sequence of circles converging to* $\mathbb{R}$.

Theorem 7.8.15 (Namazi-Souto). *Consider the sequence sequence $(M_n, \rho_{n,\epsilon})$, where $M_n, \rho_{n,\epsilon}$ are as in Theorem 7.8.11, in the Gromov-Hausdorff topology. Choose any set of base points for this sequence. The limit of any convergent subsequence of this sequence is a hyperbolic 3-manifold homeomorphic to either V_1, V_2, or $S \times \mathbb{R}$, where $S = \partial_+ V_i$.*

The following result concerns fundamental topological questions:

Theorem 7.8.16 (Namazi-Souto). *Let V_1, V_2, S, h be as above. If n is sufficiently large, then S is the unique minimal genus Heegaard surface for $V_1 \cup_{h_n} V_2$.*

There are several other complexes that are more or less related to the curve complex. The most immediate is the arc complex:

Definition 7.8.17. Let S be a compact surface with non-empty boundary. The *arc complex* of S, denoted by $A(S)$, is defined analogously to the curve complex: 0-simplices are isotopy classes of essential arcs and closed curves. Edges correspond to distinct arcs and curves, etc.

The arc complex serves as a natural generalization of the curve complex to the setting of surfaces with non-empty boundary. However, the usual curve complex is also considered in the context of surfaces with non-empty boundary with the caveat that vertices correspond to essential non-peripheral (i.e., not boundary parallel) simple closed curves. That is the complex used by Masur and Minsky in their notion of subsurface projection.

Definition 7.8.18. Suppose that X is a subsurface of S. The *subsurface projection* map $\pi : C(S) \to C(X)$ assigns to vertices of $C(S)$ simplices of $C(X)$ as follows: Let α be a simple closed curve in S that represents a vertex of $C(X)$. Denote the collection of non-peripheral curves in the boundary of a regular neighborhood of $\alpha \cup \partial X$ by α_X. The components of α_X are disjoint, so α_X corresponds to a simplex $\sigma(\alpha_X)$ in $C(X)$. We set $\pi(\alpha) = \sigma(\alpha_X)$.

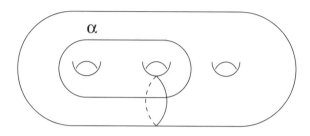

Figure 7.26. *Before the projection.*

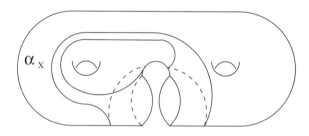

Figure 7.27. *After the projection.*

It is interesting to observe how distances behave under this projection map. In general, distance can go up under projection, but Masur and Minsky provide a characterization of this phenomenon. S. Schleimer cleverly exploits this characterization in his proof of the theorem below concerning the "end of the curve complex"; see [**140**].

Definition 7.8.19. Let X be a metric space and let $B = B(x, R) \subset X$ be a closed metric R-ball. We define the number $c(B)$ to be the number of unbounded components of $X \setminus B$. Fix a point $x \in X$. The number of *ends* of a metric space X is

$$\epsilon(X) = \sup_{R \geq 0} c(B(x, R)).$$

Note that $\epsilon(X)$ is independent of x.

Theorem 7.8.20 (Schleimer)**.** *Let S be a surface of genus at least 2 with exactly one boundary component. Then $C(S)$ has only one end.*

Recall the notion of pants decomposition from Section 2.3. Given a compact orientable surface, A. Hatcher constructed a 2-dimensional simplicial

7.8. Through the Looking Glass

complex $\mathcal{P}(S)$ as follows:

- The vertices of $\mathcal{P}(S)$ are the equivalence classes of pants decompositions of S.
- The edges of $\mathcal{P}(S)$ correspond to two types of alterations of pants decompositions of S in which one curve is replaced by another. See Figures 7.28 and 7.29.
- The 2-simplices of $\mathcal{P}(S)$ correspond to a finite set of possible types of alterations (that we will not discuss explicitly here) involving three or five curves or certain combinations of moves as in Figures 7.28 and 7.29.

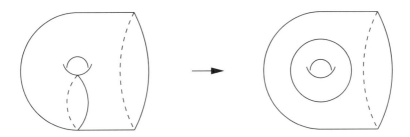

Figure 7.28. *The S-move.*

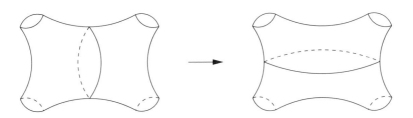

Figure 7.29. *The A-move.*

Theorem 7.8.21 (Hatcher)**.** *The pants complex is simply connected.*

See http://www.math.cornell.edu/ hatcher/Papers/pantsdecomp.pdf.

Remark 7.8.22. The pants complex is related to the dual of the curve complex.

Exercises

Exercise 1. Show that a Heegaard splitting is reducible if and only if its distance is 0. Furthermore, show that it is weakly reducible if and only if its distance is at most 1.

Exercise 2. Suppose that M is a toroidal 3-manifold. Show that the distance of any Heegaard splitting $M = V \cup_S W$ is at most 2.

Exercise 3. Show that if S is an incompressible surface in (closed orientable surface) $\times I$ with $\partial S \subset$ (closed orientable surface) $\times \{1\}$, then S is boundary parallel.

Exercise 4. Exhibit an example of a 3-manifold with infinite mapping class group.

Exercise 5. To contrast with Theorem 7.8.16, exhibit 3-manifolds with distinct minimal genus Heegaard splittings and explain which hypotheses are not satisfied.

Appendix A

General Position

The numerous technicalities involved in describing and discussing general position in the TOP category often stand in the way of clarity. We opt to discuss the analogous concept from the DIFF category. This is the notion of *transversality*. In our discussion we will need several basics from the DIFF category. We will barely touch on the key issues. For more details, see [**100**], [**48**], or [**128**].

In this appendix we consider only manifolds embedded in Euclidean space. The following theorem tells us that this in no way limits our scope:

Theorem A.0.1 (Whitney). *Every compact n-dimensional manifold admits an embedding into $\mathbb{R}^{2n}$.*

Calculus teaches us how to find the tangent space to a manifold X embedded in Euclidean space at a point x in X: It is the best linear approximation to the manifold at the given point and is denoted by $T_x(X)$. See Figures A.1 and A.2.

Fact 1. The tangent space to an n-manifold at a given point is isomorphic to $\mathbb{R}^n$.

Fact 2. The collection of all tangent spaces at points in the n-manifold X forms an $\mathbb{R}^n$-bundle over X.

Definition A.0.2. The bundle described above is called the *tangent bundle* of X and is denoted by $T(X)$.

Given a differentiable map $f : X \to Y$ and a point $x \in X$, its derivative at x, df_x, maps $T_x(X)$ to $T_{f(x)}(Y)$. Recall that a point $x \in X$ is critical if df_x fails to be surjective. A value $y \in Y$ is critical if its preimage contains

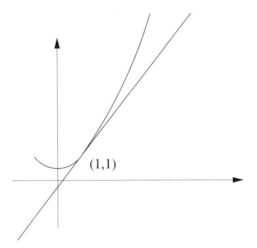

Figure A.1. *The tangent line to a curve at the point $(1,1)$.*

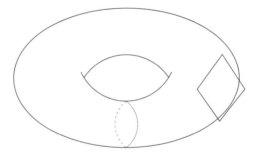

Figure A.2. *A portion of the tangent plane to a torus at a specified point.*

critical points. The following theorem lies at the heart of transversality:

Theorem A.0.3 (Sard). *The set of critical values of a differentiable map from one manifold to another has Lebesgue measure 0.*

We provide some illustrative examples:

Example A.0.4. The inclusion map $\iota : \mathbb{R} \to \mathbb{R}^2$ from $\mathbb{R}$ onto the y-axis has as its critical values the y-axis, a set of Lebesgue measure 0 in $\mathbb{R}^2$.

Example A.0.5. The map $f : \mathbb{R} \to \mathbb{R}$ given by $f(x) = x^2$ has exactly one critical point, the point 0. This point is mapped to the value 0; hence 0 is the only critical value. This is a set of measure 0 in $\mathbb{R}$.

Example A.0.6. The projection map $\pi_1 : \mathbb{R}^2 \to \mathbb{R}$ from $\mathbb{R}^2$ onto the x-axis has no critical values.

A. General Position

Example A.0.7. The map $f : \mathbb{R}^2 \to \mathbb{R}^2$ given by $f(x, y) = (x^2 y, y)$ has as its critical points the two coordinate axes. These are mapped to the values $(0, y)$, that is, the y-axis. This is a set of measure 0 in $\mathbb{R}^2$.

We now define the notion of transversality, first for submanifolds of a single manifold and then in more generality. Transversality provides the basis for many arguments in the theory of 3-manifolds. It guarantees nice intersections of submanifolds. For instance, every time we assume that a curve and a surface in a 3-manifold intersect in points (rather than, say, line segments) we are invoking general position/transversality. Note that for Y a submanifold of X and p a point in $Y \subset X$, $T_p(Y)$ is a subspace of $T_p(X)$.

Definition A.0.8 (Transversality for submanifolds). Two submanifolds Y, Z of the manifold X are *transverse* at the point $p \in Y \cap Z$ if $T_p(Y)$ and $T_p(Z)$ together span $T_p(X)$. Two submanifolds of X are *transverse* if they are transverse at each point of intersection. We write $Y \pitchfork Z$.

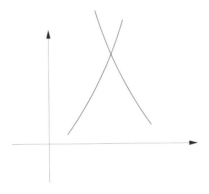

Figure A.3. *A transverse intersection of two 1-manifolds in $\mathbb{R}^2$.*

Example A.0.9. In $\mathbb{R}^2$, the x-axis and the y-axis are transverse.

Non-Example. In $\mathbb{R}^3$, the x-axis and the y-axis are not transverse.

Example A.0.10. In $\mathbb{R}^3$, the x-axis and the line parallel to the y-axis passing through $(1, 0, 1)$ are transverse.

Example A.0.11. In $\mathbb{R}^3$, the xy-plane and the z-axis are transverse.

Non-Example. In $\mathbb{R}^3$, the xy-plane and the y-axis are not transverse.

Non-Example. In $\mathbb{R}^4$, the xy-plane and the z-axis are not transverse.

Example A.0.12. In $\mathbb{R}^3$, the xy-plane and the xz-plane are transverse.

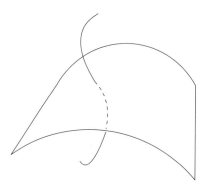

Figure A.4. *A transverse intersection of a 1-manifold and a 2-manifold in $\mathbb{R}^3$.*

Non-Example. In $\mathbb{R}^2$, the graph of the function $f : \mathbb{R} \to \mathbb{R}$ given by $f(x) = x^3$ and the x-axis are not transverse.

More generally, transversality can be formulated for maps. This formulation may appear unnatural at first glance; but it allows us to build a coherent theory.

Definition A.0.13 (Transversality in general). Let $f : Z \to X$ be a differentiable map and let Y be a submanifold of X. Let $y \in Y$ and let $z \in Z$ be such that $f(z) = y$. Then f is *transverse* to Y at z if the image of df_z and $T_y(Y)$ together span $T_y(X)$. The map f is *transverse* to Y if for every point $z \in Z$ such that $f(z) \in Y$, f is transverse to Y at z.

This condition is sometimes expressed as

$$\text{Image}(df_z) + T_y(Y) = T_y(X),$$

but note that this equality must be interpreted in the appropriate sense.

Remark A.0.14. Transversality for submanifolds is a special case of transversality in general. Indeed, two submanifolds of X are transverse if the inclusion map of one submanifold is transverse to the other submanifold.

Remark A.0.15. Let X be a manifold of dimension n_x with submanifolds Y, Z of dimensions n_y, n_z. If $Y \pitchfork Z$, then it is a consequence of the definitions that the dimension of $Y \cap Z$ is $n_y + n_z - n_x$. More generally, for $f : Z \to X$ a differentiable map transverse to $Y \subset X$, the codimension of $f^{-1}(Y)$ in Z is the same as the codimension of Y in X.

Example A.0.16. Two 1-dimensional submanifolds of a 3-manifold must be disjoint. (The proof is left as an exercise.)

Example A.0.17. A 1-dimensional submanifold and a 2-dimensional submanifold of a 3-manifold intersect in points.

Example A.0.18. Two 2-dimensional submanifolds of a 3-manifold intersect in a 1-dimensional submanifold.

Theorem A.0.19 below tells us that transversality is attainable via small isotopies called *perturbations*.

Theorem A.0.19 (The Transversality Homotopy Theorem). *For any smooth map $f : Z \to X$ and any submanifold Y of X there exists a smooth map $g : Z \to X$ homotopic to f that is transverse to Y. Moreover, suppose that $H : Z \times I \to X$ is a homotopy between f and g. Then $\forall \epsilon > 0$, the map $g_\epsilon : Z \to X$ defined by $g_\epsilon(z) = H(z, \epsilon)$ is also transverse to Y.*

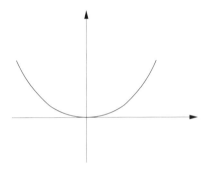

Figure A.5. *A non-transverse intersection for the x-axis and a curve.*

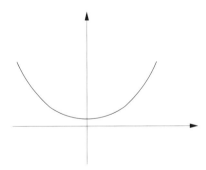

Figure A.6. *A perturbation resulting in transverse intersection.*

Recall the non-examples to transversality mentioned above: (a) In $\mathbb{R}^3$, the x-axis and the y-axis are not transverse, but the x-axis and a small perturbation of the y-axis (for instance, a translation onto a parallel axis) are transverse; (b) in $\mathbb{R}^3$, the xy-plane and the y-axis are not transverse, but the xy-plane and a small perturbation of the y-axis (for instance, a

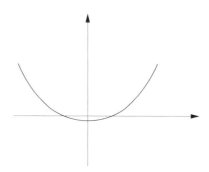

Figure A.7. *Another perturbation resulting in transverse intersection.*

translation onto a parallel axis not lying in the xy-plane) are transverse; (c) in $\mathbb{R}^4$, the xy-plane and the z-axis are not transverse, but the xy-plane and a small perturbation of the z-axis (for instance, a translation onto an axis not lying in the xy-plane) are transverse; (d) in $\mathbb{R}^2$, the graph of the function $f : \mathbb{R} \to \mathbb{R}$ given by $f(x) = x^3$ and the x-axis are not transverse, but a translation of the graph (one that raises the point of inflection of the graph above or below the x-axis) and the x-axis are transverse.

Remark A.0.20. In fact, in the Transversality Homotopy Theorem, Z is allowed to have boundary. In this case the conclusion can be strengthened: Denote the restriction of the function $h : Z \to X$ to ∂Z by ∂h. Then we can conclude that, aside from being homotopic to f and ∂f, respectively, both g and ∂g are transverse to Y.

Manifolds are metrizable spaces. Thus if Y is a submanifold of X, we can look at all points within a distance ϵ of Y. The following theorem gives us an explicit description of this set.

Theorem A.0.21 (The ϵ-Neighborhood Theorem). *Let Y be a compact k-dimensional submanifold of the n-manifold X. Let Y^ϵ denote the set of all points in X with distance less than ϵ from Y. If ϵ is sufficiently small, then Y^ϵ is a $\mathbb{B}^{n-k}$-bundle over Y.*

For a proof of this theorem, see [48]. This theorem reminds us that ϵ-neighborhoods exist and it explicitly describes them as bundles.

Remark A.0.22. The term *general position* is the TOP category and TRIANG category equivalent of the DIFF category term transversality. The term *regular neighborhood* is the TOP category equivalent of the DIFF category term ϵ-neighborhood.

Figure A.8. *An ϵ-neighborhood of a 1-manifold in a 3-manifold.*

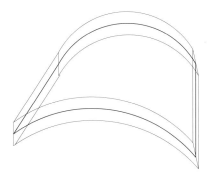

Figure A.9. *An ϵ-neighborhood of a surface in a 3-manifold.*

Theorem A.0.23 (Uniqueness of Regular Neighborhoods). *Let N_1, N_2 be regular neighborhoods of Y in X. There exists an isotopy of X fixed on Y carrying N_1 onto N_2.*

For a proof of this theorem, see [**128**].

Exercises

Exercise 1. Show that transverse 1-dimensional submanifolds of a 3-manifold must be disjoint.

Exercise 2. Draw and imagine several examples of transverse and non-transverse intersections.

Appendix B

Morse Functions

Morse functions provide for descriptions of manifolds in terms of certain basic building blocks. Moreover, they exist on all smooth manifolds. We will describe the 1-, 2-, and 3-dimensional scenarios. For more details, see [**99**] and [**48**].

Definition B.0.1. Let $f : \mathbb{R}^n \to \mathbb{R}$ be a smooth function. A critical point x of f is *non-degenerate* if the Hessian matrix of second partial derivatives is non-singular at x.

Let M be a smooth n-manifold and let $f : M \to \mathbb{R}$ be a smooth function. A point $x \in M$ is a *critical point* of f if there is a chart ϕ_α near x such that $\phi_\alpha(x) = 0$ and 0 is a critical point of $f \circ \phi_\alpha^{-1}$. It is *non-degenerate* if 0 is a non-degenerate critical point of $f \circ \phi_\alpha^{-1}$. A *critical value* of f is a value c such that $f^{-1}(c)$ contains at least one critical point.

In fact, the above definition is independent of charts. If a point in M is a critical point for one chart, then it is a critical point for all charts. Likewise, if a point in M is a non-degenerate critical point for one chart, then it is a non-degenerate critical point for all charts. As it turns out, non-degenerate critical points can be described rather informatively in terms of local coordinates; namely, if x is a non-degenerate critical point of $f : M \to \mathbb{R}$, then there are local coordinates in which

$$f(x_1, \ldots, x_n) = -\sum_{i=1}^{k} x_i^2 + \sum_{i=k+1}^{n} x_i^2.$$

This result is known as the Morse Lemma. The number k is called the *index* of the critical point.

Definition B.0.2. Let M be a manifold. A smooth proper map $f : M \to \mathbb{R}$ is a *Morse function* if all of its critical points are non-degenerate and distinct critical points correspond to distinct critical values.

Theorem B.0.3. *If M is a smooth manifold, then there exist Morse functions on M.*

In fact, for any smooth manifold M, Morse functions are dense in $C^\infty(M)$. (Here $C^\infty(M)$ is the set of smooth real-valued functions on M topologized via the notion of C^∞-convergence on compacts.) Theorem B.0.3 is a consequence of several properties of smooth manifolds including Sard's Theorem. For details, see [48].

A Morse function $f : M \to \mathbb{R}$ decomposes the manifold M into regular and singular level sets. If there are no critical values between two regular values, then their level sets are diffeomorphic. Furthermore, the regular values that lie strictly between two consecutive critical values provide a product structure on the corresponding level sets. More precisely,

$$f^{-1}((a,b))$$

is diffeomorphic to

$$f^{-1}(c) \times (a,b)$$

whenever (a,b) is an interval containing no critical values and $c \in (a,b)$. Hence to understand the manifold, we need to understand how level sets change as we pass through a critical value. Specifically, we wish to observe how the submanifold $f^{-1}((-\infty, y])$ changes as we pass through a critical value y_0. The expression of f in terms of local coordinates provides a complete picture of this change. We will consider the scenario in dimensions 1, 2, and 3.

Consider a Morse function on the closed 1-dimensional manifold $\mathbb{S}^1$. The regular level sets are discrete; hence the regular values that lie strictly between two consecutive critical values correspond to arcs in the manifold. A critical point of index 0 is modeled on x^2. Thus a new point appears and and breaks into two as we move from level sets below to level sets above the critical point. A critical point of index 1 is modeled on $-x^2$. Thus two points of $f^{-1}(y)$ merge into one as we move from level sets below to level sets above the critical point.

For a Morse function on a closed 2-dimensional manifold, the regular level sets are disjoint unions of circles. Hence preimages of intervals containing no critical values are cylinders in M. A critical point of index 0 is modeled on $x_1^2 + x_2^2$. Thus a point appears and expands into a circle as we move from level sets below to those above the critical point. A critical point of index 1 is modeled on $-x_1^2 + x_2^2$ and is also known as a *saddle singularity*.

B. Morse Functions

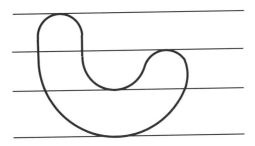

Figure B.1. *A Morse function on the circle with critical levels indicated.*

One of two things happens as we move from level sets below to those above the critical value: Either one circle is pinched and splits into two circles or two circles are wedged at a point and merge into one circle. A critical point of index 2 is modeled on $-x_1^2 - x_2^2$. Thus a circle is collapsed into a point and disappears.

Figure B.2. *A Morse function on the torus with critical levels indicated.*

For a Morse function on a closed 3-dimensional manifold, the regular level sets are surfaces. Hence the regular values that lie strictly between two consecutive critical values correspond to products of the form (surface)$\times I$. A critical point of index 0 is modeled on $x_1^2+x_2^2+x_3^2$. Thus a point appears and expands into a 2-sphere as we move from level sets below to level sets above the critical value. A critical point of index 1 is modeled on $-x_1^2 + x_2^2 + x_3^2$. Thus the two branches of a hyperboloid of two sheets meet at two points and merge into a hyperboloid of one sheet as we move from level sets below to level sets above the critical value. A critical point of index 2 is modeled on $-x_1^2 - x_2^2 + x_3^2$. Thus a hyperboloid of one sheet collapses by pinching along a circle and breaks into a hyperboloid of two sheets as we move from level sets below to level sets above the critical value. See Figure B.3.

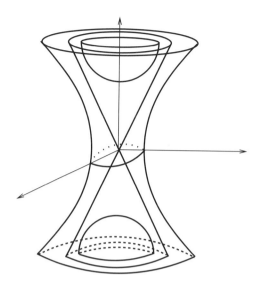

Figure B.3. *A hyperboloid of two sheets collapsing to a hyperboloid of one sheet or vice versa (x_1 is the vertical coordinate).*

A critical point of index 3 is modeled on $-x_1^2 - x_2^2 - x_3^2$. Thus a 2-sphere collapses to a point and disappears as we move from level sets below to level sets above the critical value.

Figure B.4. *Level surface below an index 1 critical value.*

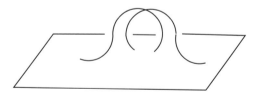

Figure B.5. *Level surface at an index 1 critical value.*

More generally, let M be a closed n-manifold, $f : M \to \mathbb{R}$ a Morse function, and $\mathbf{x}_0$ a critical point of f. The critical point $\mathbf{x}_0$ has a neighborhood homeomorphic to $[-1, 1]^n$. By the Morse Lemma there is a $k \in \mathbb{N}$ such that there are local coordinates about $\mathbf{x}_0$ in which

$$f(x_1, \ldots, x_n) = -\sum_{i=1}^{k} x_i^2 + \sum_{i=k+1}^{n} x_i^2.$$

B. Morse Functions

Figure B.6. *Level surface above an index 1 critical value.*

Let y_0 be the critical value corresponding to $\mathbf{x}_0$. Observe that the homeomorphism type of $f^{-1}((-\infty, y])$ changes as we pass through y_0. Specifically, for a, b such that $a < y_0 < b$ and such that y_0 is the only critical value between a and b, $f^{-1}((-\infty, b])$ is homeomorphic to the result of attaching a copy of $[-1,1]^n$ to $\partial f^{-1}((-\infty, a])$ along $\partial[-1,1]^k \times [-1,1]^{n-k}$. (In fact, the word "homeomorphism" can be replaced with the word "diffeomorphism" in the discussion above, though this is harder to see.)

Definition B.0.4. For M_a an n-manifold with boundary, a *k-handle* is $[-1,1]^n$ attached to M_a along $\partial[-1,1]^k \times [-1,1]^{n-k}$.

It follows that every closed manifold can be built from handles starting with a 0-handle. For instance, an n-sphere can be built from one 0-handle and one n-handle. (Simply attach the n-handle to the boundary of the 0-handle.) More generally, to build a given manifold M, choose a Morse function $h : M \to \mathbb{R}$ and then mimic the growth of $h^{-1}((-\infty, y])$ by attaching a k-handle when y passes through a critical value of index k.

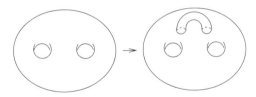

Figure B.7. *Attaching a handle of index 1 to a 3-manifold.*

Remark B.0.5. Given a Morse function $h : M \to \mathbb{R}$ on the n-manifold M, the function $-h : M \to \mathbb{R}$ is also a Morse function. Critical points of index k for h correspond to critical points of index $n - k$ for $-h$. This means that in the context of n-manifolds, a k-handle is dual to an $(n-k)$-handle in the following sense: If we can build an n-manifold M by attaching handles $h_1, h_2, \ldots, h_l$, then we can also build M by attaching these handles "upside down" and in reverse order. I.e., for each k-handle h_i, rather than attaching the k-handle h_i to the submanifold $h^{-1}((-\infty, y_0 - \epsilon])$ of M, we instead attach an $(n-k)$-handle to $h^{-1}([y_0 + \epsilon, \infty))$.

Note that a 0-handle is attached to nothing and hence is a copy of $[-1,1]^n$ appearing out of thin air. Dually, an n-handle is attached along $\partial[-1,1]^n$ and hence is a topological n-ball that caps off a spherical boundary component.

Definition B.0.6. A description of a manifold M in terms of handles is called a *handle decomposition*.

Conversely, under the right smoothness assumptions, a handle decomposition corresponds to a Morse function.

Definition B.0.7. A Morse function $f : M \to \mathbb{R}$ is *self-indexing* if, for all k, critical points of index k correspond to smaller critical values than critical points of index $k+1$.

Let M be a closed 3-manifold and let $h : M \to \mathbb{R}$ be a self-indexing Morse function. Consider a regular value r such that all critical points of h of index 0 and 1 occur below r and all critical points of index 2 and 3 occur above r. Then $f^{-1}((-\infty, r])$ is constructed from 0-handles and 1-handles and is hence a handlebody. Dually, $f^{-1}([r, \infty))$ is also a handlebody. Hence:

Remark B.0.8. A self-indexing Morse function defines a Heegaard splitting.

Conversely, a Heegaard splitting can be used to define a Morse function, though the Morse function is not unique.

Exercises

Exercise 1. Suppose that a closed n-manifold M admits a Morse function with exactly two critical points. Prove that M is homeomorphic to $\mathbb{S}^n$. (In general, M need not be diffeomorphic to $\mathbb{S}^n$. This fact was used by Milnor in [98] to construct examples of 7-dimensional topological spheres with exotic smooth structures.)

Exercise 2. Suppose that a closed 3-manifold M admits a Morse function with exactly 4 critical points. Prove that M is a lens space or $\mathbb{S}^3$.

Exercise 3. Explore how Morse functions interact with other structures on manifolds. For example, read [58].

Bibliography

1. Colin C. Adams, *The knot book. An elementary introduction to the mathematical theory of knots*, W. H. Freeman and Company, New York, 1994. Reprinted with corrections, Amer. Math. Soc., Providence, RI, 2004.

2. Ian Agol, *The Virtual Haken Conjecture*, arXiv:1204.2810 [math.GT].

3. Ian Agol and Yi Liu, *Presentation length and Simon's conjecture*, J. Amer. Math. Soc. **25** (2012), no. 1, 151–187. MR2833481

4. Lars Ahlfors and Leo Sario, *Riemann surfaces*, Princeton Mathematical Series, vol. 26, Princeton University Press, Princeton, NJ, 1960.

5. Selman Akbulut, *Scharlemann's manifold is standard*, Ann. of Math. (2) **149** (1999), no. 2, 497–510. MR1689337 (2000d:57033)

6. James Waddell Alexander, *Note on Riemann spaces*, Bull. Amer. Math. Soc. **26** (1920), no. 8, 370–372.

7. _____, *A lemma on systems of knotted curves*, Proc. Nat. Acad. Sci. U.S.A. **9** (1923), no. 3, 93–95.

8. Riccardo Benedetti and Carlo Petronio, *Lectures on hyperbolic geometry*, Universitext, Springer-Verlag, Berlin, 1992. MR1219310 (94e:57015)

9. Laurent Bessières, Gérard Besson, Sylvain Maillot, Michel Boileau, and Joan Porti, *Geometrisation of 3-manifolds*, EMS Tracts in Mathematics, vol. 13, European Mathematical Society (EMS), Zürich, 2010. MR2683385 (2012d:57027)

10. Joan S. Birman, *Braids, links, and mapping class groups*, Princeton University Press, Princeton, NJ, 1974, Annals of Mathematics Studies, No. 82. MR0375281 (51 #11477)

11. Steven Bleiler and Martin Scharlemann, *A projective plane in $\mathbf{R}^4$ with three critical points is standard. Strongly invertible knots have property P*, Topology **27** (1988), no. 4, 519–540.

12. Michel Boileau, Donald Collins, and Heiner Zieschang, *Scindements de Heegaard des petites variétes de Seifert*, C. R. Acad. Sci. Paris Sér. I Math. **305** (1989), no. 12, 557–560.

13. Michel Boileau and Jean-Pierre Otal, *Sur les scindements de Heegaard du tore T^3*, J. Differential Geom. **32** (1990), no. 1, 209–233.

14. Francis Bonahon and Jean-Pierre Otal, *Scindements de Heegaard des espaces lenticulaires*, C. R. Acad. Sci. Paris Sér. I Math. **294** (1982), no. 17, 585–587.

15. Brian H. Bowditch, *Intersection numbers and the hyperbolicity of the curve complex*, J. Reine Angew. Math. **598** (2006), 105–129.

16. Mark Brittenham and Ying-Qing Wu, *The classification of exceptional Dehn surgeries on 2-bridge knots*, Comm. Anal. Geom. **9** (2001), no. 1, 97–113. MR1807953 (2001m:57008)

17. Jeffrey F. Brock, Kenneth W. Bromberg, Richard D. Canary, and Yair N. Minsky, *Local topology in deformation spaces of hyperbolic 3-manifolds*, Geom. Topol. **15** (2011), no. 2, 1169–1224. MR2831259

18. M. Brown, *A proof of the generalized Schoenflies Theorem*, Bull. Amer. Math. Soc. **66** (1960), 74–76.

19. Gerhard Burde, Heiner Zieschang, and Michael Heusener, *Knots*, third ed., de Gruyter Studies in Mathematics, vol. 5, Walter de Gruyter & Co., Berlin, 2013.

20. Danny Calegari and David Gabai, *Shrinkwrapping and the taming of hyperbolic 3-manifolds*, J. Amer. Math. Soc. **19** (2006), no. 2, 385–446.

21. Alberto Candel and Lawrence Conlon, *Foliations. I*, Graduate Studies in Mathematics, vol. 23, American Mathematical Society, Providence, RI, 2000. MR1732868 (2002f:57058)

22. _____, *Foliations. II*, Graduate Studies in Mathematics, vol. 60, American Mathematical Society, Providence, RI, 2003. MR1994394 (2004e:57034)

23. Huai-Dong Cao and Xi-Ping Zhu, *A complete proof of the Poincaré and Geometrization Conjectures—application of the Hamilton-Perelman Theory of the Ricci flow*, Asian Journal of Mathematics **10** (2006), no. 2, 165–492.

24. A. J. Casson and C. McA. Gordon, *A loop theorem for duality spaces and fibred ribbon knots*, Invent. Math. **74** (1983), no. 1, 119–137.

25. Andrew J. Casson and Steven A. Bleiler, *Automorphisms of surfaces after Nielsen and Thurston*, London Mathematical Society Student Texts, vol. 9, Cambridge University Press, Cambridge, 1988. MR964685 (89k:57025)

26. Jean Cerf, *La stratification naturelle des espaces de fonction différentiables réelles et le théorème de la pseudo-isotopie*, Inst. Hautes Études Sci. Publ. Math. **39** (1970), 5–173.

27. Éric Charpentier, Étienne Ghys, and Annick Lesne (eds.), *The scientific legacy of Poincaré*, History of Mathematics, vol. 36, American Mathematical Society, Providence, RI, 2010, Translated from the 2006 French original by Joshua Bowman. MR2605614 (2011b:00005)

28. J. H. Conway, *An enumeration of knots and links, and some of their algebraic properties*, Computational Problems in Abstract Algebra (Proc. Conf., Oxford, 1967), Pergamon, Oxford, 1970, pp. 329–358. MR0258014 (41 #2661)

29. Daryl Cooper and Martin Scharlemann, *The structure of a solvmanifold's Heegaard splittings*, Proceedings of 6th Gökova Geometry-Topology Conference, vol. 23, 1999, pp. 1–18. MR1701636 (2000h:57034)

30. Marc Culler, C. McA. Gordon, J. Luecke, and Peter B. Shalen, *Dehn surgery on knots*, Ann. of Math. (2) **125** (1987), no. 2, 237–300.

31. Oliver T. Dasbach and Tao Li, *Property P for knots admitting certain Gabai disks*, Topology Appl. **142** (2004), no. 1-3, 113–129.

32. M. Dehn, *Über die Topologie des dreidimensionalen Raumes*, Math. Ann. **69** (1910), 137–168.

33. _____, *Die Gruppe der Abbildungsklassen. (German) Das arithmetische Feld auf Flächen*, Acta Math. **69** (1938), no. 1, 135–206.

34. Charles Delman and Rachel Roberts, *Alternating knots satisfy strong Property P*, Comment. Math. Helv. **74** (1999), no. 3, 376–397.

35. Y. Diao, J. C. Nardo, and Y. Sun, *Global knotting in equilateral random polygons*, JKTR **10** (2001), no. 4, 597–697.

36. Yuanan Diao, Nicholas Pippenger, and De Witt Sumners, *On random knots. Random knotting and linking*, JKTR **3** (1994), no. 3, 419–429.

37. Manfredo Perdigão do Carmo, *Riemannian geometry*, Mathematics: Theory & Applications, Birkhäuser Boston Inc., Boston, MA, 1992, Translated from the second Portuguese edition by Francis Flaherty. MR1138207 (92i:53001)

38. Cornelia Drutu and Michael Kapovich, *Lectures on geometric group theory*, preprint.

39. A. Fathi, F. Laudenbach, and V. Poenaru, *Travaux de Thurston sur les surfaces*, Astérisque **66-67** (1979), 1–284.

40. M. Feighn, *Branched covers according to J. W. Alexander*, Collect. Math. **37** (1986), no. 1, 55–60.

41. David Gabai, *Foliations and the topology of 3-manifolds*, J. Differential Geom. **18** (1983), no. 3, 445–503. MR723813 (86a:57009)

42. _____, *Genera of the arborescent links*, Mem. Amer. Math. Soc., vol. 59, American Mathematical Society, Providence, RI, 1986.

43. _____, *Foliations and the topology of 3-manifolds. III*, J. Differential Geom. **26** (1987), no. 3, 479–536. MR910018 (89a:57014b)

44. _____, *3 lectures on foliations and laminations on 3-manifolds*, Laminations and foliations in dynamics, geometry and topology (Stony Brook, NY, 1998), Contemp. Math., vol. 269, Amer. Math. Soc., Providence, RI, 2001, pp. 87–109. MR1810537 (2002g:57032)

45. David Gabai and Ulrich Oertel, *Essential laminations in 3-manifolds*, Ann. of Math. (2) **130** (1989), no. 1, 41–73. MR1005607 (90h:57012)

46. C. McA. Gordon and J. Luecke, *Knots are determined by their complements*, Bull. Amer. Math. Soc. (N.S.) **20** (1989), no. 1.

47. V.K.A.M. Gugenheim, *Piecewise linear isotopy and embedding of elements and spheres. I, II*, Proc. LMS **3** (1953), 29–53.

48. V. Guillemin and A Pollack, *Differential topology*, Prentice Hall, Inc., Englewood Cliffs, NJ, 1974.

49. André Haefliger, *Variétés feuilletées*, Ann. Scuola Norm. Sup. Pisa (3) **16** (1962), 367–397. MR0189060 (32 #6487)

50. _____, *Feuilletages riemanniens*, Astérisque (1989), no. 177-178, Exp. No. 707, 183–197, Séminaire Bourbaki, Vol. 1988/89. MR1040573 (91e:57047)

51. _____, *Travaux de Novikov sur les feuilletages*, Séminaire Bourbaki, Vol. 10, Soc. Math. France, Paris, 1995, Exp. No. 339, 433–444. MR1610457

52. Wolfgang Haken, *Ein Verfahren zur Aufspaltung einer 3-Mannigfaltigkeit in irreduzible 3-Mannigfaltigkeiten*, Math. Z. **76** (1961), 427–467. MR0141108 (25 #4519c)

53. _____, *Theorie der Normalflächen*, Acta Math. **105** (1961), 245–375. MR0141106 (25 #4519a)

54. Richard S. Hamilton, *A compactness property for solutions of the Ricci flow*, Amer. J. Math. **117** (1995), no. 3, 545–572. MR1333936 (96c:53056)

55. _____, *The formation of singularities in the Ricci flow*, Surveys in differential geometry, Vol. II (Cambridge, MA, 1993), Int. Press, Cambridge, MA, 1995, pp. 7–136. MR1375255 (97e:53075)

56. Kevin Hartshorn, *Heegaard splittings of Haken manifolds have bounded distance*, Pacific J. Math. **204** (2002), no. 1, 61–75. MR1905192 (2003a:57037)

57. William James Harvey, *Geometric structure of surface mapping class groups*, Homological group theory (Proc. Sympos., Durham, 1977), London Math. Soc. Lecture Note Ser., vol. 36, Cambridge Univ. Press, Cambridge-New York, 1979, pp. 255–269.

58. Joel Hass, Paul Norbury, and J. Hyam Rubinstein, *Minimal spheres of arbitrarily high Morse index*, Comm. Anal. Geom. **11** (2003), no. 3, 425–439. MR2015753 (2004i:53082)

59. Allen Hatcher, *Notes on basic 3-manifold topology*, preprint.

60. P. Heegaard, *Sur l'"Analysis situs"*, Bull. Soc. Math. France **44** (1916), 161–242. MR1504754

61. Geoffrey Hemion, *The classification of knots and 3-dimensional spaces*, Oxford Science Publications, The Clarendon Press Oxford University Press, New York, 1992. MR1211184 (94g:57015)

62. John Hempel, *3-manifolds as viewed from the curve complex*, Topology **40** (2001), no. 3, 631–657.

63. _____, *3-manifolds*, AMS Chelsea Publishing, Providence, RI, 2004, reprint of the 1976 original.

64. H. Hilden, M. T. Lozano, and J. M. Montesinos, *Universal knots*, Bull. Amer. Math. Soc. (N.S.) **8** (1983), no. 3, 449–450.

65. Dale Husemoller, *Fibre bundles*, third ed., Graduate Texts in Mathematics, vol. 20, Springer-Verlag, New York, 1994. MR1249482 (94k:55001)

66. Michael Hutchings, *Floer homology of families. I*, Algebr. Geom. Topol. **8** (2008), no. 1, 435–492. MR2443235 (2009h:57046)

67. William Jaco, *Lectures on three-manifold topology*, CBMS Regional Conference Series in Mathematics, vol. 43, American Mathematical Society, Providence, RI, 1980.

68. William Jaco and Ulrich Oertel, *An algorithm to decide if a 3-manifold is a Haken manifold*, Topology **23** (1984), no. 2, 195–209.

69. William Jaco and J. Hyam Rubinstein, *PL minimal surfaces in 3-manifolds*, J. Differential Geom. **27** (1988), no. 3, 493–524.

70. _____, *PL equivariant surgery and invariant decompositions of 3-manifolds*, Adv. in Math. **73** (1989), no. 2, 149–191.

71. William Jaco and Peter B. Shalen, *Seifert fibered spaces in irreducible, sufficiently-large 3-manifolds*, Bull. Amer. Math. Soc. **82** (1976), no. 5, 765–767. MR0415623 (54 #3706)

72. Klaus Johannson, *On exotic homotopy equivalences of 3-manifolds*, Geometric topology (Proc. Georgia Topology Conf., Athens, GA, 1977), Academic Press, New York, 1979, pp. 101–111. MR537729 (80m:57002)

73. _____, *Topology and combinatorics of 3-manifolds*, Lecture Notes in Mathematics, vol. 1599, Springer-Verlag, Berlin, 1995. MR1439249 (98c:57014)

74. Michael Kapovich, *Hyperbolic manifolds and discrete groups*, Modern Birkhäuser Classics, Birkhäuser Boston Inc., Boston, MA, 2009, reprint of the 2001 edition. MR2553578 (2010k:57039)

75. Svetlana Katok, *Fuchsian groups*, Chicago Lectures in Mathematics, University of Chicago Press, Chicago, IL, 1992. MR1177168 (93d:20088)

76. Mikhail Khovanov, *Link homology and categorification*, International Congress of Mathematicians. Vol. II, Eur. Math. Soc., Zürich, 2006, pp. 989–999. MR2275632 (2008f:57009)

77. Robion Kirby, *A calculus for framed links in S^3*, Invent. Math. **45** (1978), no. 1, 35–56. MR0467753 (57 #7605)

78. Robion Kirby and Paul Melvin, *The E_8-manifold, singular fibers and handlebody decompositions*, Proceedings of the Kirbyfest (Berkeley, CA, 1998), Geom. Topol. Monogr., vol. 2, Geom. Topol. Publ., Coventry, 1999, pp. 233–258. MR1734411 (2000j:57051)

79. Bruce Kleiner and John Lott, *Notes on Perelman's papers*, Geom. Topol. **12** (2008), no. 5, 2587–2855. MR2460872 (2010h:53098)

80. Helmut Kneser, *Geschlossene Flächen in dreidimensionalen Mannigfaltigkeiten*, Jber. d. D. Math. Verein. **38** (1929), 248–260.

81. Tsuyoshi Kobayashi, *A construction of arbitrarily high degeneration of tunnel numbers of knots under connected sum*, J. Knot Theory Ramifications **3** (1994), no. 2, 179–186. MR1279920 (95g:57011)

82. P. B. Kronheimer and T. S. Mrowka, *Witten's conjecture and property P*, Geom. Topol. **8** (2004), 295–310 (electronic). MR2023280 (2004m:57023)

83. Marc Lackenby, *Heegaard splittings, the virtually Haken conjecture and property (τ)*, Invent. Math. **164** (2006), no. 2, 317–359. MR2218779 (2007c:57030)

84. Tao Li, *Heegaard surfaces and measured laminations. I. The Waldhausen conjecture*, Invent. Math. **167** (2007), no. 1, 135–177. MR2264807 (2008h:57033)

85. _____, *An algorithm to determine the Heegaard genus of a 3-manifold*, Geom. Topol. **15** (2011), no. 2, 1029–1106. MR2821570 (2012m:57033)

86. _____, *Rank and genus of 3-manifolds*, J. Amer. Math. Soc. **26** (2013), no. 3, 777–829. MR3037787

87. W. B. Raymond Lickorish, *An introduction to knot theory*, Graduate Texts in Mathematics, vol. 175, Springer-Verlag, New York, 1997.

88. W. B. R. Lickorish, *A representation of orientable combinatorial 3-manifolds*, Ann. of Math. **76** (1962), no. 2, 531–540.

89. Charles Livingston, *Knot theory*, Carus Mathematical Monographs, vol. 24, Mathematical Association of America, Washington, DC, 1993. MR1253070 (94m:57021)

90. D. D. Long, A. Lubotzky, and A. W. Reid, *Heegaard genus and property τ for hyperbolic 3-manifolds*, J. Topol. **1** (2008), no. 1, 152–158. MR2365655 (2008j:57036)

91. Feng Luo, *Automorphisms of the complex of curves*, Topology **39** (2000), no. 2, 283–298. MR1722024 (2000j:57045)

92. Jason Manning, *Algorithmic detection and description of hyperbolic structures on closed 3-manifolds with solvable word problem*, Geom. Topol. **6** (2002), 1–26.

93. William S. Massey, *Algebraic topology: An introduction*, Springer-Verlag, New York, 1977, reprint of the 1967 edition, Graduate Texts in Mathematics, Vol. 56. MR0448331 (56 #6638)

94. H. A. Masur and Y. N. Minsky, *Geometry of the complex of curves. I. Hyperbolicity*, Invent. Math. **138** (1999), no. 1, 103–149.

95. _____, *Geometry of the complex of curves. II. Hierarchical structure*, Geom. Funct. Anal. **10** (2000), no. 4, 902–974.

96. Sergei Matveev, *Algorithmic topology and classification of 3-manifolds*, Algorithms and Computation in Mathematics, vol. 9, Springer-Verlag, Berlin, 2003. MR1997069 (2004i:57026)

97. Curtis T. McMullen, *The evolution of geometric structures on 3-manifolds*, Bull. Amer. Math. Soc. (N.S.) **48** (2011), no. 2, 259–274. MR2774092 (2012a:57024)

98. John Milnor, *On manifolds homeomorphic to the 7-sphere*, Ann. of Math. (2) **64** (1956), 399–405. MR0082103 (18,498d)

99. _____, *Morse theory. Based on lecture notes by M. Spivak and R. Wells*, Annals of Mathematics Studies, vol. 51, Princeton University Press, Princeton, NJ, 1963.

100. John W. Milnor, *Topology from the differentiable viewpoint*, Princeton Landmarks in Mathematics, Princeton University Press, Princeton, NJ, 1997, based on notes by David W. Weaver, revised reprint of the 1965 original. MR1487640 (98h:57051)

101. Edwin E. Moise, *Affine structures in 3-manifolds. V. The triangulation theorem and Hauptvermutung*, Ann. of Math. **56** (1952), no. 2, 96–114.

102. _____, *Geometric topology in dimensions 2 and 3*, Springer-Verlag, New York, 1977, Graduate Texts in Mathematics, Vol. 47. MR0488059 (58 #7631)

103. Frank Morgan, *Manifolds with density and Perelman's proof of the Poincaré conjecture*, Amer. Math. Monthly **116** (2009), no. 2, 134–142. MR2478057 (2010a:53164)

104. John Morgan and Gang Tian, *Ricci flow and the Poincaré conjecture*, Clay Mathematics Monographs, vol. 3, American Mathematical Society, Providence, RI, 2007. MR2334563 (2008d:57020)

105. John W. Morgan and Hyman Bass (eds.), *The Smith conjecture*, Pure and Applied Mathematics, vol. 112, Academic Press Inc., Orlando, FL, 1984, Papers presented at the symposium held at Columbia University, New York, 1979. MR758459 (86i:57002)

106. Yoav Moriah and Jennifer Schultens, *Irreducible Heegaard splittings of Seifert fibered spaces are either vertical or horizontal*, Topology **37** (1998), no. 5, 1089–1112. MR1650355 (99g:57021)

107. Kanji Morimoto, *There are knots whose tunnel numbers go down under connected sum*, Proc. Amer. Math. Soc. **123** (1995), no. 11, 3527–3532. MR1317043 (96a:57022)

108. Kanji Morimoto, Makoto Sakuma, and Yoshiyuki Yokota, *Examples of tunnel number one knots which have the property "1 + 1 = 3"*, Math. Proc. Cambridge Philos. Soc. **119** (1996), no. 1, 113–118. MR1356163 (96i:57007)

109. James R. Munkres, *Topology: A first course*, Prentice-Hall Inc., Englewood Cliffs, NJ, 1975. MR0464128 (57 #4063)

110. Kunio Murasugi, *On the braid index of alternating links*, Trans. Amer. Math. Soc. **326** (1991), no. 1, 237–260.

111. Hossein Namazi, *Heegaard splittings and hyperbolic geometry*, ProQuest LLC, Ann Arbor, MI, 2005, Thesis (Ph.D.)–State University of New York at Stony Brook. MR2707466

112. Hossein Namazi and Juan Souto, *Heegaard splittings and pseudo-Anosov maps*, Geom. Funct. Anal. **19** (2009), no. 4, 1195–1228. MR2570321 (2011a:57035)

113. Y. Ni, *Uniqueness of PL minimal surfaces*, Acta Math. Sin. (Engl. Ser.) **23** (2007), no. 6, 961–964.

114. S. P. Novikov, *The topology of foliations*, Trudy Moskov. Mat. Obšč. **14** (1965), 248–278. MR0200938 (34 #824)

115. Peter Ozsváth, András I. Stipsicz, and Zoltán Szabó, *Combinatorial Heegaard Floer homology and nice Heegaard diagrams*, Adv. Math. **231** (2012), no. 1, 102–171. MR2935385

116. Peter Ozsváth and Zoltán Szabó, *Knot Floer homology and the four-ball genus*, Geom. Topol. **7** (2003), 615–639. MR2026543 (2004i:57036)

117. _____, *Knot Floer homology, genus bounds, and mutation*, Topology Appl. **141** (2004), no. 1-3, 59–85. MR2058681 (2005b:57028)

118. Peter S. Ozsváth and Zoltán Szabó, *Knot Floer homology and rational surgeries*, Algebr. Geom. Topol. **11** (2011), no. 1, 1–68. MR2764036 (2012h:57056)

119. C. D. Papakyriakopoulos, *On Dehn's lemma and the asphericity of knots*, Ann. of Math. **66** (1957), no. II, 1–26.

120. R. C. Penner and J. L. Harer, *Combinatorics of train tracks*, Annals of Mathematics Studies, vol. 125, Princeton University Press, Princeton, NJ, 1992. MR1144770 (94b:57018)

121. G. Perelman, *The entropy formula for the Ricci flow and its geometric applications*, http://arxiv.org/abs/math/0211159.

122. _____, *Finite extinction time for the solutions to the Ricci flow on certain three-manifolds*, http://arxiv.org/abs/math/0307245.

123. _____, *Ricci flow with surgery on three-manifolds*, http://arxiv.org/abs/math/0303109.

124. John G. Radcliffe, *Foundations of hyperbolic manifolds*, second ed., Graduate Texts in Mathematics, vol. 149, Springer Verlag, New York, 2006.

125. Yo'av Rieck and Eric Sedgwick, *Thin position for a connected sum of small knots*, Algebr. Geom. Topol. **2** (2002), 297–309 (electronic).

126. Igor Rivin, *Euclidean structures on simplicial surfaces and hyperbolic volume*, Ann. of Math. (2) **139** (1994), no. 3, 553–580. MR1283870 (96h:57010)

127. Dale Rolfsen, *Knots and links*, Mathematics Lecture Series, vol. 7, Publish or Perish Inc., Houston, TX, 1990, corrected reprint of the 1976 original.

128. C. Rourke and B. Sanderson, *Introduction to piecewise-linear topology*, Springer Study Edition, Springer-Verlag, Berlin-New York, 1982, reprint of the 1972 original.

129. Hyam Rubinstein and Martin Scharlemann, *Comparing Heegaard splittings of non-Haken 3-manifolds*, Topology **35** (1996), no. 4, 1005–1026. MR1404921 (97j:57021)

130. _____, *Transverse Heegaard splittings*, Michigan Math. J. **44** (1997), no. 1, 69–83. MR1439669 (98c:57017)

131. _____, *Comparing Heegaard splittings—the bounded case*, Trans. Amer. Math. Soc. **350** (1998), no. 2, 689–715. MR1401528 (98d:57033)

132. Martin Scharlemann, *Constructing strange manifolds with the dodecahedral space*, Duke Math. J. **43** (1976), no. 1, 33–40. MR0402760 (53 #6574)

133. _____, *Heegaard splittings of compact 3-manifolds*, Handbook of geometric topology, North-Holland, Amsterdam, 2002, pp. 921–953. MR1886684 (2002m:57027)

134. Martin Scharlemann and Jennifer Schultens, *The tunnel number of the sum of n knots is at least n*, Topology **38** (1999), no. 2, 265–270. MR1660345 (2000b:57013)

135. _____, *Annuli in generalized Heegaard splittings and degeneration of tunnel number*, Math. Ann. **317** (2000), no. 4, 783–820.

136. Martin Scharlemann and Abigail Thompson, *Thin position and Heegaard splittings of the 3-sphere*, J. Differential Geom. **39** (1994), no. 2, 343–357.

137. _____, *Thin position for 3-manifolds*, Geometric topology (Haifa, 1992), Contemp. Math., vol. 164, Amer. Math. Soc., Providence, RI, 1994, pp. 231–238.

138. Martin G. Scharlemann, *Unknotting number one knots are prime*, Invent. Math. **82** (1985), no. 1, 37–55. MR808108 (86m:57010)

139. Saul Schleimer, *Waldhausen's theorem*, Workshop on Heegaard Splittings, Geom. Topol. Monogr., vol. 12, Geom. Topol. Publ., Coventry, 2007, pp. 299–317. MR2408252

140. _____, *The end of the curve complex*, Groups Geom. Dyn. **5** (2011), no. 1, 169–176. MR2763783

141. Horst Schubert, *Knoten und Vollringe*, Acta Math. **90** (1953), 131–286.

142. _____, *Über eine numerische Knoteninvariante*, Math. Z. **61** (1954), 245–288.

143. Jennifer Schultens, *The classification of Heegaard splittings for (compact orientable surface)* $\times S^1$, Proc. London Math. Soc. (3) **67** (1993), no. 2, 425–448.

144. Jennifer Schultens and Richard Weidman, *On the geometric and the algebraic rank of graph manifolds*, Pacific J. Math. **231** (2007), no. 2, 481–510. MR2346507 (2009a:57030)

145. Peter Scott, *The geometries of 3-manifolds*, Bull. London Math. Soc. **15** (1983), no. 5, 401–487. MR705527 (84m:57009)

146. Herbert Seifert and William Threlfall, *Seifert and Threlfall: A textbook of topology*, Pure and Applied Mathematics, vol. 89, Academic Press Inc. [Harcourt Brace Jovanovich Publishers], New York, 1980, translated from the German edition of 1934 by Michael A. Goldman, with a preface by Joan S. Birman, with "Topology of 3-dimensional fibered spaces" by Seifert, translated from the German by Wolfgang Heil. MR575168 (82b:55001)

147. Zlil Sela, *The isomorphism problem for hyperbolic groups I*, Ann. Math. **141** (1995), 217–283.

148. I. M. Singer and J. A. Thorpe, *Lecture notes on elementary topology and geometry*, second ed., Undergraduate Texts in Mathematics, Springer-Verlag, New York-Heidelberg-Tokyo, 1976.

149. John Stallings, *On the loop theorem*, Ann. of Math. (2) **72** (1960), 12–19. MR0121796 (22 #12526)

150. Norman Steenrod, *The topology of fibre bundles*, Princeton Landmarks in Mathematics, Princeton University Press, Princeton, NJ, 1999, reprint of the 1957 edition, Princeton Paperbacks. MR1688579 (2000a:55001)

151. Abigail Thompson, *Thin position and bridge number for knots in the 3-sphere*, Topology **36** (1997), no. 2, 505–507.

152. _____, *Algorithmic recognition of 3-manifolds*, Bull. Amer. Math. Soc. (N.S.) **35** (1998), no. 1, 57–66.

153. William P. Thurston, *The geometry and topology of three-manifolds*, http://www.msri.org/publications/books/gt3m/.

154. _____, *Three-dimensional geometry and topology*, vol. 1, Princeton Mathematical Series, vol. 35, Princeton University Press, Princeton, NJ, 1997.

155. F. Waldhausen, *Heegaard-Zerlegungen der 3-Sphäre*, Topology **7** (1968), no. 2, 195–203.

156. Friedhelm Waldhausen, *On irreducible 3-manifolds which are sufficiently large*, Ann. of Math. (2) **87** (1968), 56–88. MR0224099 (36 #7146)

Index

2π-Theorem, 237
2-bridge, 119
2-fold branched cover, 141
3-colorable, 107
$S_1 + S_2$, 157
δ-thin, 251
k-handle, 175, 273
k-simplex, 13
n-manifold, 1
n-torus, 2
r-skeleton, 143

Agol, 237
Alexander Trick, 45, 53
Alexander's Theorem, 67, 71
almost normal, 170
alternating, 115
anannular, 224
Anosov, 246
arational, 246
arborescent knot, 119
arc complex, 257
atlas, 1
atoroidal, 96, 224

barycentric coordinates, 144
base space, 58
bicollar, 60
blackboard framing, 227
boundary, 4
boundary incompressible, 74
boundary irreducible, 70
braid, 122

branch locus, 43
branched covering, 43
bridge number, 129
bundle, 57
bundle atlas, 58
bundle chart, 58

characteristic submanifold, 98
chart, 1
classification of surfaces, 37
closure of a braid, 123
cocore, 176
collar, 60
companion, 121
complete, 218
complexity, 248
compressible, 73
compression body, 202
cone, 148
connected sum, 23, 108
convex, 219
core, 176
covering, 42
covering space, 139
crossing number, 114
curve complex, 248
cut, 63
cut and paste argument, 99
cyclic surgery, 232

decomposing annulus, 210
decomposing sphere, 167
Dehn filling, 227

Dehn surgery, 227, 232
Dehn surgery space, 237
Dehn twist, 50, 245
Dehn's Lemma, 76
Dehn's Theorem, 139
destabilization, 190
dilation, 218
dimension, 1
Diophantine system, 159
distance, 248
distance of the Heegaard splitting, 252
double, 90

embedding, 4
ends, 258
equivalent, 19, 22, 59, 102, 178, 205
essential, 32, 73, 75
Euler characteristic, 20
exceptional fiber, 87

face, 14
fiber, 58, 87
fibering, 58
fill, 253
filling, 45, 246
foliation, 238
fundamental, 160
fundamental group, 139

Gauss-Jordan elimination, 159
general position, 55, 147
generalized Heegaard splitting, 204
generic, 256
genus, 38, 112, 177, 193
geodesic, 218
geometric intersection number, 249
geometric manifold, 22
Geometrization Conjecture, 224
glue, 99
glued, 178
good system, 188
granny knot, 109
graph manifold, 97
Gromov, 237
Gromov hyperbolic, 251
Gromov-Hausdorff topology, 256

Haken 3-manifold, 75
handle decomposition, 274
handlebody, 177
Hausdorff distance, 256
Heegaard diagram, 181
Heegaard genus, 212

Heegaard splitting, 178, 202, 228
height function, 64, 128
hierarchy, 80
homotopy, 23
homotopy equivalent, 83
Hopf link, 117
horizontal, 92, 194, 254
hyperbolic arc length, 215
hyperbolic n-manifold, 221
hyperbolic volume, 215

ideal triangle, 219
incompressible, 73
independent, 162
index, 211, 269
inessential, 32
infinite cyclic cover, 139
innermost disk argument, 98
intersection number, 31
inversion, 219
irreducible, 62, 183
isometry, 22
isomorphic, 59
isomorphism, 58
isotopy, 23

Jones polynomial, 126
Jordan Curve Theorem, 30, 64

Kneser-Haken finiteness, 162
knot, 102
knot diagram, 103
knot invariant, 113

lamination, 243
leaf, 238
leaves, 243
length, 82, 149, 235
lens space, 83
link, 102, 146
linking number, 117
longitude, 106, 227
Loop Theorem, 79

mapping class group, 48, 255
measure, 152
meridian, 106, 227
meridian disks, 177
metric triangulation, 173
Möbius band, 10
Montesinos knot, 119
Morse function, 9, 64, 270
Mostow Rigidity Theorem, 225

natural framing, 227
negative curvature, 173
non-degenerate, 269
non-separating, 71
normal curve, 149
normal disk, 151
normal isotopy, 152
normal surface, 151
normal triangle, 150
nugatory, 115

open regular neighborhood, 60
opposite, 11
orientable, 10
orientation, 31
orientation-preserving, 12
orientation-reversing, 12
outermost arc argument, 99

pair of pants, 5
pants decomposition, 40
partial ordering, 144
pattern, 121
Perelman, 224
periodic, 53
PL least area, 173
Poincaré Conjecture, 84, 224, 237
Poincaré-Hopf Index Theorem, 26, 64, 67, 239, 242
pretzel knot, 119
prime, 24
prime decomposition, 162, 166
prime factorization, 55
prism manifolds, 90
projection, 58
projective measured lamination, 245
projectively equivalent, 245
proper, 73
properly discontinuously, 221
Property P, 236
pseudo-Anosov, 53, 247, 250
punctured, 162

rank, 212
real projective space, 2
reduced, 115
reducible, 53, 62, 183
Reeb foliation, 239
Reebless, 241
regular neighborhood, 50, 60
Reidemeister moves, 106
restriction, 59

Ricci flow, 224
Riemann-Hurwitz Theorem, 44
rotation, 219

satellite knot, 121
Scharlemann cycle, 235
Schönflies Theorem, 63, 67, 104, 163
section, 60
Seifert fibered space, 88, 238, 253
Seifert surface, 110, 119, 226
Seifert's Algorithm, 110
self-indexing, 274
separating, 32, 71
simple arc, 29
simple closed curve, 29
simplices, 248
simplicial complex, 15, 19
simplicial isomorphism, 19
simplicial map, 18
simply connected at infinity, 86
small, 136
sphere, 2
Sphere Theorem, 79
spine, 194
square knot, 109
square restriction, 157
stabilization, 182
stable, 247
standard Heegaard splitting, 194
standard innermost disk argument, 92
star, 146
strongly irreducible, 183, 205
subcomplex, 19
subdivision, 145
submanifold, 3
subsurface projection, 257
swallow-follow torus, 121
sweepout, 196

tangent bundle, 261
taut, 242
thick level, 131
thin level, 132
thin position, 130
torus knot, 105
total space, 58
translation, 219
transversal, 241
transversality, 102
transverse, 263, 264
transverse invariant measure, 243
transverse isotopy, 243

transversely, 123
trivial, 23
tunnel system, 208

uniquely ergodic, 246
universal, 139
unknot, 102
unknotting number, 116
unstable, 247
untelescoping, 206
upper half-space model, 217

vertical, 92, 194, 254

Waldhausen's Theorem, 83
weight, 152
weighted intersection number, 245
Whitehead manifold, 84
width, 130
wild knot, 101
Wirtinger presentation, 137

(